AF114315

THE GREAT SEA FLOODS OF 1953

THE GREAT SEA FLOODS OF 1953

The Records of P.J.O. (John) Trist

Edited by
DENISE PARKINSON

The Boydell Press

Suffolk Records Society
Volume LXVII

A Suffolk Records Society publication
First published 2024
The Boydell Press, Woodbridge

ISBN 978-1-83765-176-4
ISBN 978-1-73980-962-1 (members)

Issued to subscribing members for the year 2024

The Boydell Press is an imprint of Boydell & Brewer Ltd
PO Box 9, Woodbridge, Suffolk IP12 3DF, UK
and of Boydell & Brewer Inc.
668 Mt Hope Avenue, Rochester, NY 14620-2731, USA
website: www.boydellandbrewer.com

A catalogue record for this book is available
from the British Library

Cover image: Flood damage and temporary repair work at Shottisham Creek, River Deben estuary, Suffolk, after the tidal surge of 31 January / 1 February. Photograph taken in February 1953 by Robert Adams, Chief Engineer for the Isle of Wight River Board, drafted to Suffolk to assist with emergency repairs (SRO, K681/2/93/3). © Crown copyright

CONTENTS

ILLUSTRATIONS

Plates

Maps of Suffolk coastal lands annotated by P.J.O. Trist between pp. 120 and 121

ACKNOWLEDGEMENTS

It is little more than a year since I first looked at the collection of papers and maps that ultimately comprise this volume in the archives at The Hold in Ipswich. I offer special thanks to two people for that introduction: Harvey Osborne and Bridget Hanley. Harvey, course leader for MA History at the University of Suffolk and my academic supervisor at the time, pointed me towards Trist's *Survey of Agriculture of Suffolk* (1971) for background information on how the 1953 sea floods affected farming. The final chapter of that book is dedicated to the event and unpublished records in 'East Suffolk County Archives' are referenced in its notes.

Bridget Hanley, senior archivist at Suffolk Archives at The Hold, responded with enthusiasm to my initial approach for help with locating these unpublished records. During the editorial process she and her colleagues have helped enormously, not least in producing digital files and facilitating countless visits to the search room. I would also like to thank Suffolk County Council for its permission to reproduce the records, deposited on behalf of the long defunct County Agricultural Executive Committee.

I am grateful to the Suffolk Records Society for its interest in these important records. To its council member, Vivienne Aldous, I offer thanks for her invaluable practical advice. Finally, I am indebted to Harvey Osborne, who is also chairman of the society in addition to his role at the university, for initiating the project and guiding me through the transcription and research as my general editor. This volume would not have been completed without him.

Denise Parkinson
June 2023

ABBREVIATIONS

AEC	Agricultural Executive Committee
BPP	British Parliamentary Papers
CAEC	County Agricultural Executive Committee
CAO	County Advisory/Agricultural Officer
CLA	Country Landowners Association
COI	Central Office of Information
DAO	District Advisory Officer (several within one county)
DUKW	Amphibious transporter from General Motors, USA
DWS	Drainage and Water Supply division within MAF
EADT	*East Anglian Daily Times*, daily regional newspaper
EVW	European Volunteer Worker
FLS	Fellow of the Linnean Society of London
HMSO	His Majesty's Stationery Office
ICI	Imperial Chemical Industries
MAF	Ministry of Agriculture and Fisheries
MAFF	Ministry of Agriculture, Fisheries and Food (after 1955)
MRAC	Member of the Royal Agricultural College
NAAS	National Agricultural Advisory Service
NFU	National Farmers' Union
OD	Ordnance Datum, mean sea level
OS	Ordnance Survey
PD	Provincial Director of NAAS, Suffolk being the eastern province
Qtrs	An unsettled measurement of mass or volume of dry goods, 1 quarter being 64 gallons or 535 imperial pounds (Weights and Measures Act 1824)
RDC	Rural District Council, local government district within a county
UDC	Urban District Council, local government district within a county
WVS	Women's Voluntary Services, a welfare support charity

PREFACE

The coastal floods that occurred on the weekend of 31 January and 1 February 1953 represent the worst natural disaster that Britain experienced in the twentieth century. A combination of wind, high tide, and low air pressure caused North Sea levels to rise and surge through sea defences, ravaging over 900 miles (1450 kilometres) of coastline between Yorkshire and Kent. Over 300 people died as a direct result of the flooding, which also damaged homes, industrial facilities, and infrastructure.[1] Several comprehensive and authoritative accounts of sundry aspects of the disaster were published at the time.[2] Yet the story of the 1953 floods did not receive a great deal of attention from writers for several decades thereafter. Perhaps it was felt that there was little more to be said.

Over time, however, a growing popular interest in social history and wider anxieties stemming from an increased risk of flooding due to climate change has stimulated a renewed interest in the events of 1953. Ongoing debates about how, and to what extent, low-lying areas of England can be defended from the sea have drawn on aspects of the human tragedy in 1953 to highlight current and emerging threats to populations in areas since developed on land vulnerable to flooding from tidal surges.[3] Writers concerned with the politics of social responsibility have also questioned whether accounts of the bravery and resilience shown by survivors of the 1953 floods mask the fact that the British people were at the time inadequately protected from the risk of flooding due to a lack of foresight, expertise, and action on the part of central government.[4] As a consequence, changing perceptions of risk and responsibility have affected how the 1953 disaster has been presented at different times.

In places where significant numbers of people perished, the social cost of the disaster has often been commemorated since with public memorials, and the experiences of survivors have been recorded in local histories.[5] The anniversary of the

---

1 Across the Channel, Holland suffered profound damage and loss of life more than five times greater.

2 BPP 1953–54, XIII.511, Home Office, Scottish Office, Ministry of Housing and Local Government, Ministry of Agriculture and Fisheries, Report of the Departmental Committee on Coastal Flooding [Waverley Report]; H. Grieve, *The Great Tide: The Story of the 1953 Flood Disaster in Essex* (Chelmsford, 1959); Lord Mayor of London's National Flood and Tempest Fund, *The Sea Came In: The History of the Lord Mayor of London's National Flood & Tempest Distress Fund* (London, 1959).

3 For example, D. Summers, *The East Coast Floods* (Newton Abbott, 1978); P.J. Baxter, 'The East Coast Big Flood, 31 January–1 February 1953: A Summary of the Human Disaster', *Philosophical Transactions of The Royal Society A* No. 363 (2005), pp. 1293–312; G. O'Hara, *The Politics of Water in Post-war Britain* (London, 2017).

4 For example, M. Pollard, *North Sea Surge: The Story of the East Coast Floods of 1953* (Lavenham, 1978) ; A. Hall, 'The Rise of Blame and Recreancy in the United Kingdom: A Political and Scientific Autopsy of the North Sea Flood of 1953', *Environment and History* 17, No. 3 (2011), pp. 379–408.

5 For example, R. Flaxman, *Wall of Water: Lowestoft and Oulton Broad During the 1953 Flood* (Lowestoft, 1993); N.R. Storey, *Flood Alert! Norfolk 1953* (Stroud, 2003); P. Rennoldson-Smith, *The 1953 Essex Flood Disaster: The People's Story* (Cheltenham, 2012); S. Armstrong, and L. Easthope, '"That Night": Unlocking the Memories of Loss on Canvey Island in 1953', *International Journal of Regional and Local History* 13, No. 2 (2018), pp. 134–46.

great flood is periodically marked in televisual and print media, with the reproduction of black and white images of the event acting to highlight paradoxically both its separation from our own times and its relative closeness. As a result, a dim awareness of the 1953 sea floods has been retained in the popular imagination as the event passes from living memory into documented history. However, the uneven nature of that documented history and a paucity of primary source material relating to experiences outside major centres of population have contributed to a sometimes partial understanding of the event. Much less is known, for example, about how the floods impacted the rural landscape and sectors such as agriculture.

In fact, most of the land flooded was in agricultural use. This caused one agricultural journalist to declare that 'the weekend of January 31st–February 1st will be remembered as marking one of the most serious farming disasters of the century'.[6] This was no small matter given that food security was a national concern at the time, with many foodstuffs still subject to post-war rationing. As a result, small armies of servicemen and farm workers were swiftly mobilised in the aftermath of the floods to make emergency repairs to sea and river walls to protect not only vulnerable towns but valuable food-growing land. In the longer term, flood defence and agricultural policies combined to significantly improve the protection afforded to some farmland from coastal flooding for the first time. Furthermore, farmers reclaimed a greater area of marsh, converting it into productive arable land and pasture with government support in the form of technical advice and financial subsidies.

The documents presented here provide a rich insight into the rural and agrarian aspect of the 1953 disaster. The records left by John Trist, affectionately dubbed 'Mr. Agriculture' by the Suffolk farming community that he served for 21 years, are rare, both in terms of the source itself and the quality of the writing.[7] Each of the engaging accounts authored by this expert witness are subtly different. Trist's typescript was intended for general public interest, his diary was private, and his draft notes were for an official report to be read by fellow professionals. Therefore, the publication of this multi-faceted collection of documents presents a unique opportunity to widen access to fresh records that plug some of the gaps in the popular history of the 1953 sea floods.

Through this volume I hope that the members of the Suffolk Records Society and a wider audience with an interest in twentieth-century British history will come to appreciate aspects of the career of John Trist and his enormous contribution to public life. The accounts left by this assiduous public servant of the events that followed the weekend of 31 January and 1 February 1953 serve to powerfully enhance our understanding of both the sea floods in East Anglia and the post-war agricultural scene in Suffolk in general.

Denise Parkinson

6 'The East Coast Floods', *Agriculture* 59, No.12 (March 1953), p. 554.

7 'Suffolk's own "Mr. Agriculture" retires this month' (*EADT*, 3 February 1971).

INTRODUCTION

This volume presents a collection of documents concerning the 1953 east coast sea floods, prepared by P.J.O. (John) Trist, who was employed by the National Agricultural Advisory Service (NAAS) of the Ministry of Agriculture (MAF) as county advisory officer (CAO) for East Suffolk from 1946 until his retirement in 1971.[1] Following the 1953 sea floods, Trist planned and organised the recovery of agricultural land in East Suffolk. In June of that year, he was awarded officer of the Order of the British Empire (OBE) in recognition of his 'services during the recent floods in the Eastern Counties'.[2]

In 1962 the Ministry of Agriculture, Fisheries and Food (MAFF) compiled an internal report on the effects of the 1953 sea floods on agriculture, prepared by a working party set up by the main advisory committee on sea flooded land. Written in two parts, the report first discusses the disaster: the impact on agricultural land, emergency measures, and government assistance for farming. The second part records the work undertaken to measure, monitor, and mitigate the effects of sea flooding and to restore land to productivity. The report (which was not in public circulation) is the only record for agriculture that covers all six counties that were significantly affected by the 1953 floods.[3]

Trist was a prominent member of the working party responsible for that report. He wrote the first part and he collated national statistics for livestock and farm stock losses, and for acreages of flooded agricultural land. By way of introduction, he described the storm that disturbed the state of the sea and caused the unprecedented surge tide in 1953. He wrote that:

> The floods of February 1953, which inundated large areas of land on the east coast of Britain, were the outcome of an exceptional combination of meteorological conditions. Abnormally low barometric pressures developed off the north coast of Scotland and were centred in a depression which on the last day of January, moved in a south-easterly direction reaching Denmark by nightfall. During this time a northerly gale of record severity developed, and combined with a moderate spring tide, produced a surge of water in the North Sea which bore down on the coast of Britain and the Netherlands, breached the sea defences and invaded the coastal land. Flooding took place first in Northumberland and in the space of nine hours on the night of 31st January the surge had travelled down to Dover, overtopping or breaking through the sea walls in more than a thousand places. The damage to life and property which followed has been described as 'probably the worst peace-time disaster Great Britain has ever known'.[4]

1 By the time of his retirement, Trist had been promoted to county agricultural adviser for East and West Suffolk.

2 *Supplement to The London Gazette*, 1 June 1953, p. 2961.

3 MAF 221/11, The Effect on Agriculture of the East Coast Floods, 1953 [Draft report of the working party set up by the main advisory committee on sea flooded land, Ministry of Agriculture and Fisheries].

4 Ibid., p. 1.

Plate 1. P.J.O. Trist – Suffolk's own 'Mr. Agriculture' – photographed on the occasion of his retirement from the National Agricultural Advisory Service after 25 years as county adviser and chief officer of the agricultural executive committee. Published 3 February 1971. © EADT

This description is consistent with an account given by Professor J.A. Steers of Cambridge University to the Waverley Committee in 1953.[5] This committee examined the causes of the floods, and found that an unprecedented combination of a huge tidal surge, enormous waves created by the storm conditions, and a reasonably high spring tide, was to blame. Furthermore, it was noted that the topography of the east coast of England made it susceptible to storm damage caused by wave action. It is an area lacking in natural sea defences, with long stretches of low-lying lands protected only by dunes and shingle ridges. Parts of this coast are inundated with numerous tidal estuaries with marsh land on either side, protected only by earth embankments. In Suffolk (as in Essex) the sea coming up estuaries over-topped and breached riverbanks, flooding marshes in agricultural use and in some cases flooding coastal towns from behind.[6]

The consequential loss of life and damages to property were summarised in a civil engineering conference later the same year. Engineers tasked with finding technical solutions to the challenge of reinstating sea defences were informed that:

> The most tragic result of this flooding was the loss of more than 300 human lives. There were also material losses on a large scale: 24,000 dwelling houses were flooded and damaged, some beyond repair; about 200 major industrial premises were inundated, and in many cases were partly out of production for several weeks. Public utility services were interrupted; twelve gasworks, two large electricity generating stations, and many smaller stations and sub-stations suffered varying degrees of dislocation and damage. Water supply and sewage disposal services were interrupted, and some wells and underground water resources were contaminated with salt. Transport suffered badly. Eleven trunk or class I roads became impassable as well as innumerable secondary and unclassified roads. These … also suffered badly from heavy traffic engaged on carrying materials for the subsequent repair work. More than 200 miles of railway were put out of action, nearly 50,000 tons of filling and ballast being needed for reinstating tracks undermined by the flood waters. Last but not least, there was the agricultural damage. Agricultural land flooded by salt water accounted for 162,905 acres. Little of this area produced a worthwhile crop in 1953 and much work is having to be done to rehabilitate it … A feature of the damage was that it was not limited to simple flooding. Such was the force of the water that not only did it batter down property, but it also swept and deposited vast quantities of sand and other debris over wide areas.[7]

Flood damage to the highly productive agricultural areas close to the east coast represented a significant blow to food production at a time of continuing anxiety about food supply in post-war Britain. The gravity of the situation was outlined by

5 There was no public enquiry into the flood disaster but the Departmental Committee on Coastal Flooding (the Waverley Committee) was formed to report on causation and to recommend changes in sea defence policy. Steers, an eminent geologist who had studied the East Anglian shoreline for decades was the expert adviser to the committee. His papers about the Suffolk coast are published in three parts in the *Proceedings of Suffolk Institute of Archaeology and Natural History* 19 (1927). Part I is about the evolution of the Suffolk shore in general; Part II describes some of the changes that have taken place along the shore from Yarmouth to Aldeburgh, and Part III deals with Orford Ness.

6 J.A. Steers, 'The East Coast Floods', *The Geographical Journal* 119, No. 3 (1953), pp. 280–95; BPP 1953–54, XIII.511, Home Office, Scottish Office, Ministry of Housing and Local Government, Ministry of Agriculture and Fisheries, Report of the Departmental Committee on Coastal Flooding [the Waverley Report], pp. 5–8.

7 Institution of Civil Engineers, *Conference on the North Sea Floods of 31 January/1 February, 1953: a collection of papers presented at the Institution in December 1953* (London, 1954), pp. 5–7.

the Prime Minister, Sir Winston Churchill, in a speech that he gave on 17 February 1953 to National Farmers' Union (NFU) members in London:

> British agriculture is a national issue … It is not a question of convenience but one of life and death for ourselves and for all we mean to the freedom of the world. Fifty million people, all dwelling in a small island, growing enough food for say, thirty millions, is indeed a spectacle of majesty and insecurity, which history has not often seen before. There are therefore twenty million Britons who depend from day to day for what they eat upon their ability to sell goods and render services on and across the seas - in return for which we get the rest of our daily bread. Home-grown food of Britain must with great urgency be raised by 1956, to 60 per cent above what it was before the war. That is a conviction all Britons, in their senses, are determined to sustain … the balance between population and food supply has tilted to an uneconomic, unwholesome, and dangerous extreme. A vast growing world towers up around us and reveals increasing strain and tension. Populations almost everywhere out pace the food supply. The balance of food production in this island has got to be altered in a marked and decisive manner, and altered soon.[8]

The floods disrupted this national food campaign and concerns about how long land might be out of production were justified. There was not only potential for long-term damage to arable land and pasture, but there were also immediate concerns about lost winter crops, how to accommodate livestock, and the impact the disruption would have on milk and stock sales. All of which threatened to imperil food supplies, farm profitability, and the viability of some holdings. Food production was also threatened by the losses of a range of farming capital, including buildings, farm equipment, and stocks of feeding stuff that could not readily be made good. In recognition of this, Churchill pledged to support farmers. Within days of the disaster, he said that, 'The restoration of the land to productivity after being damaged by the salt sea demands action guided by the highest scientific and expert authorities.'[9]

Trist was one of the 'highest scientific and expert authorities' that Churchill had in mind. He was an employee of NAAS, a specialised department of MAF set up to educate and advise farmers. The department was involved in education and the research of experimental approaches to all aspects of farming, including machinery use, livestock husbandry, horticulture, and poultry and milk production. Its staff worked in close association with County Agricultural Executive Committees (CAEC), which were statutory bodies, established under the Agriculture Act of 1947, with the duty of promoting agricultural development and efficiency. This local authority provided direct assistance to farmers by maintaining contract services for pest control, labour, and land drainage, among others. The East Suffolk Agricultural Executive Committee (AEC) (as elsewhere) comprised members appointed by the MAF, nominated by the Country Landowners' Association (CLA), the NFU and the Agricultural Workers' Union to represent the interests of landowners, farmers, and farmworkers. As advisory officer, Trist combined the function of head of the

8 'Speech of the Prime Minister, the Rt. Hon. Winston Churchill, O.M., C.H., D.L., M.P., at the Annual Dinner of the National Farmers' Union on February 17, 1953', *Agriculture* 59, No. 12 (March 1953), pp. 551–54.

9 Ibid., p. 551.

advisory service in the county and chief officer of the AEC providing advice, and shaping and enforcing agricultural policy.[10]

In 1953, recent changes in land drainage policy meant it was not particularly easy to demarcate the activities of the AECs and the newly formed river boards.[11] Hilda Grieve wrote that in Essex,

> generally, the agricultural executive committee was concerned with improvements to under-drainage of the land on each farm, by means of tile or mole drains, which remove excess sub-surface water through a system of underground drainage or soil channels, and ditching to remove excess standing water. For this purpose, the committee maintained and controlled a drainage department of its own with plant and equipment, technical staff and a regular supply of labour. But once water reached a 'common watercourse', of which the blockage, pollution, diversion or flooding would cause a public nuisance, it came under the jurisdiction of the river board.[12]

This was also the arrangement in Suffolk. In the emergency response to the sea floods, the interests of these two authorities came together and the river board and the farming community shared two urgent objectives: first, to close the breaches in the sea and river defences to prevent further inundation on the next high tide; second, to rid the land of seawater.

The records Trist left in the archive provide a unique insight into the collaboration between East Suffolk AEC and the East Suffolk and Norfolk River Board. They reveal that a multitude of problems arose. From making emergency repairs to river walls and pumping seawater off the land, to rescuing livestock and taking the first steps in rehabilitating the land. Trist records the practical measures adopted to deal with difficulties, which included a lack of experienced engineers and ground workers, shortage of the proper tools and heavy machinery, trouble accessing remote river walls on marshes, and reliance on a few suitable foremen to oversee a large civilian (sometimes unskilled and sometimes unwilling) work force that was scrambled to the sea and river walls in the 'battle of the breaches'.[13]

The records also highlight a continuation of the wartime policy that as much food as possible should be grown at home, and that, so far as achievable, every acre of potentially good land should be cultivated. Contemporary news reports illustrate the attitudes of landowners and officials to this policy, and Trist provides examples of how it had been applied on some farms in coastal Suffolk since the end of the war. In all, Trist names more than 400 farmers and farm workers and the places they

10 A discussion of the background to the formation of NAAS in 1946, including concerns of the possibility of conflicts between advisory and statutory parts of county adviser jobs, is provided in P. Brassley, D. Harvey, M. Lobley, and M. Winter, *The Real Agricultural Revolution: The Transformation of British Farming 1939–1985* (Woodbridge, 2021), pp. 62–7. Trist describes the history of NAAS in Suffolk: *A Survey of the Agriculture of Suffolk* (London, 1971), pp. 303–16.

11 River boards, established by the 1948 River Boards Act, were authorities that controlled land drainage, fisheries, and river pollution and had other functions relating to rivers, streams, and inland waters in England and Wales between 1950 and 1965. River boards were accountable to the land drainage and water supplies division of the MAF. Prior to this, catchment boards (since 1930) had responsibility for each of the main rivers in England and Wales, replacing the 400-year-old precept (regulated by the Statute of Sewers) that only those who directly benefitted from drainage works paid for them.

12 H. Grieve, *The Great Tide: The Story of the 1953 Flood Disaster in Essex* (Chelmsford, 1959), p. 470.

13 'Battle of the breaches', a phrase adopted in the media to describe the race to close gaps in sea defences in the two weeks following the flood, to prevent further flooding on the next spring tide.

Plate 2. A gang of men repairing flood damage to the river wall on the Ramsholt Dock marshes, alongside the River Deben between the Ramsholt Arms pub and the church, after the sea floods. © Robert Simper

Plate 3. John Trist on the river wall at Gedgrave marshes, near Orford, while an excavator and a bulldozer work at filling in a gap behind an emergency sandbag defence erected by volunteer workers. Photograph published 27 February 1953. © EADT

lived and worked. This provides a partial but unique insight into who was farming in the flooded areas of East Suffolk and how their lands and livestock were affected by flooding. He also records the part played by farmers and farm workers in the emergency rescues of livestock, voluntary provision of farm labour for wall repairs, and work done to restore the land to productivity.

Philip John Owen Trist (1908–96)

Trist, the son of a doctor in general practice, spent most of his childhood in Cornwall and Somerset, where he was schooled at King's College, Taunton.[14] He gained his first practical farming experience after he left school, spending three tough years on sheep farms in New Zealand before returning to study at the Royal Agricultural College in Cirencester.[15] Two cash-strapped years of farming in Herefordshire followed, before Trist joined the Wiltshire Agricultural Accounting Society as an advisory officer in 1935, reading for an economics degree at the same time.[16] The outbreak of war in 1939 led to a brief spell in the Royal Artillery before a return to agriculture. For most of the war Trist served on the Essex War AEC. His task in Essex was to oversee the reclamation of vast tracts of derelict land and the conversion of many acres of grassland to cereal crops to increase domestic food production, an experience that would serve him well in his organisation of the response to the 1953 flood disaster.[17]

One of the more remarkable features of the collection of documents presented in this volume is Trist's description of the flood damage he surveyed during a light aircraft flight on 4 February 1953. This flight along the flooded coastline from the River Stour, which forms the boundary between Suffolk and Essex, to the Waveney, on the Suffolk–Norfolk border, caused Trist to paint in words a vivid picture of the landscape. For example, when describing one of the deluged areas that he observed from the air he wrote:

> Poor Minsmere! Once again she lay drowned in a sea of salt … below us in the centre I saw the dead body of an area of marsh which only the previous year I had restored to health … it was a child which I had nurtured with much patience and care – and it lay dead below me.[18]

The heartfelt anguish Trist expressed is remarkable. Surveying and report-writing were fundamental requisites for an agricultural adviser, but the value of the documents presented here is their uniquely personal insight that goes beyond the perfunctory prose of formal reports.

As a man from the ministry, Trist was a professional advocate for modern farming methods including the introduction of machinery that was still relatively new to agricultural practice in East Anglia. Describing the immense task of moving mud and debris deposited by the sea, he writes enthusiastically about the benefits of bulldozers, and DUKWs, military relics from the war pressed into service. However,

---

14 He was born in Putney Bridge Road, London on 22 April 1908, but the family soon moved to Cornwall. King's College was a boarding school for boys aged 8–18.

15 He graduated as a Member of the Royal Agricultural College (MRAC) in 1932.

16 He graduated BA (Hons) from Bristol University in 1939.

17 See Richard Trist (his son) 'Bibliography of P.J.O. Trist, OBE, BA, MRAC, FLS, Agriculturalist and Botanist (1908–1996)', *Suffolk Natural History* 39 (2003), p. 76.

18 'The great sea floods of 1953', below, p. 9.

it is also evident that Trist harboured a deep respect for the highly skilled, often manual, craft traditions that had long been an integral part of agricultural and land management work, but which he often perceived as being lost or usurped. On one occasion when he sat in his local pub, captivated by an old-hand explaining time-honoured methods of pest control for the benefit of maintaining secure river walls, Trist reflected that, 'Illustration and experience was retold whilst I listened intently: and I also thought of the practices of the past which in our modern age are being lost to the cause of speed and the paramount demagogue of economics.'[19]

Trist was a countryman. His observations on the weather and his glee at the prospect that the high winds would offer the opportunity of a challenging shot as he crouched in his pigeon hide are testament to this. He documented the effect of the flood on wild animals, from worms to rabbits, and he listed the myriad birds he observed along the coast. Characteristically, a description of a scene that caught his attention on his way home from Woodbridge one evening is imbued with affection for the natural world. Trist wrote that,

> gulls were dipping over the water, until their formation was broken up by the clumsy flapping of a cormorant which took off in an upstream direction. As I left the wall I heard the whine of powerful wings and three swans passed over in line ahead and made towards the old tide mill.[20]

Perhaps not altogether surprisingly for a man who was ultimately an employee of a large bureaucratic department of the state Trist frequently expressed an impatience with officialdom. Yet he often displayed due deference to authority and powerful figures in the local establishment. He also demonstrated a sympathy for landowners and respect for workers in equal measure. He had a particular appreciation for those who could endure hard manual work. Trist's admiration for those volunteers who came to assist with the wall repairs is marked by his comment that, 'each man played his part; and there was that obvious fellow feeling between the farm worker and the land which supported him and his family'.[21]

Fortitude and stoicism are virtues that were celebrated in contemporary narratives of the flood disaster. This has caused some to suggest that hackneyed ideas about a national resolve and resilience (with roots in the wartime 'Blitz spirit') were marshalled to disguise the awful reality that lives and livelihoods were blighted by a disaster that could have been avoided. In this regard, Trist's comments about those 'volunteer' wall workers, who were not always the most enthusiastic recruits, is revealing. Only within his notes for the internal MAFF report did he disclose that,

> the employment of unemployed men from the Ministry of Labour was in most cases a travesty of work … there were complaints of pay, cold and … there were also agitators who encouraged men to leave the job … it is unlikely that many of this class of labour are worth considering on similar occasions.[22]

Trist appears to have happily subjugated self-interest to do his duty (and more) during the emergency and he was disappointed not to find that same sense of duty and social responsibility in others.

19 'Records in the history of the Suffolk coastline', below, p. 56.
20 'The great sea floods of 1953', below, p. 22.
21 Ibid. p. 19.
22 'Sea floods 1953 – Report', below, p. 108.

Trist has been described as a 'knowledgeable, meticulous and kind man' by those who knew him well, and they have suggested that the sea floods marked a turning point in his career.[23] His work on salt-tolerant grasses was the beginning of his serious study of botany.[24] There is certainly evidence of his emerging interest in plants in the records transcribed in this volume. Trist writes of the marram grass used to create 'bent hills' as a natural defence against sea erosion. Historic planting on the sea wall at Minsmere also gets his attention. In a characteristic display of local knowledge, Trist noted that, 'this wall, known to thousands who have had tea and bathed off the shingle beach below the sandy cliff, was topped with tamarisk originally planted by Mr. Rope's father, who preceded his son in the occupation of the Lower Abbey farm'.[25]

The floods may have marked a shift in Trist's interests; they also heralded a shift in approaches to the management of marginal coastal lands in Suffolk. Trist successfully argued that the damage caused to many thousands of acres of marshes in the county presented the opportunity to 'introduce a new management policy designed to tap the hitherto wasted potential of this area'.[26] As a consequence, areas of rough grazing were restored and water-logged areas were reclaimed to bring the land into cultivation for the first time. This caused Trist to reflect that, to his mind, 'the flood of 1953 was the finest thing which ever happened to the Suffolk marshes. The defences are raised, the land has been levelled, the ditches restored and the installation of new pumping stations …' transforming the area into productive agricultural use.[27]

In 1971 the publication of Trist's *Survey of the Agriculture of Suffolk* for the Royal Agricultural Society of England coincided with both the abolition of the CAECs and Trist's retirement from the MAFF.[28] He went on to become the first conservation officer with the Suffolk Trust for Nature Conservation.[29] Trist continued writing for the rest of his life, with many works reflecting his specialist knowledge of grasses and grasslands.[30] His published bibliography lists more than 130 written works and also documents that his personal records of the 1953 floods remained unpublished, until now that is.[31]

23 A.C. Copping, 'Philip John Owen Trist OBE, BA, MRAC, FLS 1908–1996', *Suffolk Natural History* 33 (1997), pp. 100–01.

24 Derek Wells, 'John Trist (1908–1996)', *Nature in Cambridgeshire* 39 (1997), p. 82.

25 'The great sea floods of 1953', below, **p. 10**.

26 P.J.O. Trist, 'Suffolk Marsh Reclamation Policy', *Agriculture* 61, No. 7 (November 1954), pp. 328–32; Trist outlined that for centuries these marshes had been neglected because of a constant threat of tidal flooding, which created a condition under which no useful grasses could survive. For pasture the marshes were of poor quality. The uneven land, puddled with saltwater in the winter and parched in the summer, was overrun with salt-tolerant species that inhibited the growth of diverse herbage.

27 *A Survey of the Agriculture of Suffolk* (London, 1971), p. 339.

28 3 February 1971, *EADT.*

29 Later known as Suffolk Wildlife Trust. Trist was active in establishing several new reserves, writing the first guide to Suffolk reserves, and later acting as warden for some of the Breckland reserves.

30 His collection of grasses was left to Cambridge University herbarium.

31 Richard Trist, 'Bibliography of P.J.O. Trist, OBE, BA, MRAC, FLS, Agriculturalist and Botanist (1908–1996),' *Suffolk Natural History* 39 (2003), pp. 75–84.

The documents

This volume presents transcriptions of the previously unpublished records of John Trist relating to the 1953 coastal floods and their impact on agriculture in East Suffolk. The records in question were originally created between 1953 and 1957 and all are currently held in the Ipswich branch of the Suffolk Record Office, now called Suffolk Archives. They comprise a typescript of two distinct chapters, a personal account of the 1953 floods entitled 'The great sea floods of 1953 (as it affected East Suffolk)', and a further account entitled 'Records in the history of the Suffolk coastline', with coverage extending back to the sixteenth century (A2727/1/1); Trist's notes and observations from February to March 1953, recorded in diary form (A2727/1/2); and, lastly, draft notes that Trist prepared for an official MAFF report into the effects of the floods on agriculture (A2727/1/3). Also published here are images of 16 maps, which were individually annotated by Trist to illustrate both the extent of inundation by the sea in 1953 and the dimensions of every breach in the flood defences along the coastline and rivers of East Suffolk (A2727/2/1–16).[32]

It is perhaps curious that Trist never published the materials he assembled on the 1953 floods and the history of flooding and coastal engineering along the Suffolk coastline himself, especially given that he had considerable experience and pedigree as an author. His first book, on the topic of land reclamation during the Second World War, had been published six years prior, and he was a frequent contributor of articles about farming matters for the local press. He also wrote for *Agriculture,* the monthly journal for farmers published by MAFF.[33] And in April 1953 the Central Office of Information (COI) commissioned him to write an article about the treatment of sea-damaged farmlands and the challenges involved in restoring them to fertility. According to a letter from the Overseas Press Section, Trist drafted a thousand words, for which the COI agreed to pay him eight guineas.[34]

There are features of his typescript account of the 1953 floods and history of the Suffolk coastline that suggest that Trist possibly intended it for publication as part of a larger piece of work at some point. The page numbering begins at 69, and the chapters were originally numbered four and five, although this was later swapped round, then crossed out. Also, the front page of the typescript is endorsed with the words: 'Written by P.J.O. Trist, OBE, BA, MRAC, County Agricultural Officer, who reserves all publication rights …'. However, there is nothing in the archives to suggest the existence of a larger work, apart from the report for MAFF. It is therefore only possible to speculate why Trist did not independently pursue publication, although he may have been deterred by an awareness that other officially sanctioned accounts of the floods were in development in East Anglia by 1954.[35]

32 These documents form part of a larger collection of materials in Suffolk Archives, donated by Trist and his family (A2727 and HD285).

33 For example, Trist, 'Suffolk Marsh Reclamation Policy', *Agriculture* 61, No. 7, November 1954, pp. 328–32; and 'County Reports on The East Coast Sea Floods', *Agriculture* 60, No. 1, April 1953, p. 24.

34 Letter from Victoria Chapelle, Features Section, to G.C. Johnson, MAF, dated 24 April 1953 (MAF 220/9, Correspondence between East Suffolk AEC and MAF Floods Emergency Division, Whitehall).

35 *The Great Tide* was commissioned by Essex County Council shortly after the flood, with the intention of documenting the complete story of the disaster. It was written and researched by Hilda Grieve, then senior assistant archivist at the Essex Record Office. Published in 1959, the book has since been described by the writer Ken Worpole as 'one of the great works of twentieth century English social history' ('The Great Tide Remembered', Essex Record Office website). *The Sea Came In* was the official record of the Lord Mayor's Flood and Tempest Distress Fund, also commissioned shortly

Trist placed the documents he produced in the aftermath of the floods in the county archive for safe keeping relatively soon after they were written. The records presented in this volume are taken from two accessions of materials that Trist donated, in 1954 and 1957, on behalf of the CAEC to what was at the time East Suffolk County Council. Trist's decision to transfer his work into the custody of the local authority for posterity at a relatively early stage was probably influenced by his recent experience of a minor disaster. On 2 February 1953, Trist recorded that, while attending an annual dinner organised by the Beccles branch of the NFU a year previously, he had received alarming news, which had caused him to leave the dinner early.[36] There was a fire at Fern Hill, Melton, the 24-roomed Georgian country house which had for some years been the headquarters of the East Suffolk AEC. The property was gutted on the night of Monday 28 January 1952, and it was reported that 'only a small part of the valuable office equipment was salvaged, and many records including all the legal documents of the committee had been destroyed'. Clearly this affected Trist. When interviewed at the time he said: 'we can still go on with our work, but it is an awful blow losing the work of years'. Indeed, as a result of the fire Trist had 'lost his personal note-book containing records of over 25 years, and his personal library of books'.[37] This misfortune, rather than a loss of interest in seeing his story in print, may have shaped Trist's subsequent attitude to his carefully constructed documentary account of the floods.

The records that Trist bequeathed to future generations are the closest that Suffolk has to the archive meticulously assembled by Hilda Grieve for the neighbouring county of Essex and published in the form of *The Great Tide* by Essex County Council in 1954. In comparison with Grieve's work, Trist's account of the floods is brief. It is also intentionally selective, with a focus on issues aligned to Trist's professional concerns. Trist chose to highlight his personal experience of the disaster and its aftermath in parts of rural East Suffolk and to focus on later remedial work conducted on the coastal marshes. This reflected the deliberate and conscious focus of the author. As he himself said, 'my concern and responsibility at the time was directed towards the land and I had little first hand knowledge of what went on in the flooded coastal towns'.[38]

Considerable thought has been given to the order in which to present the documents in this volume. Strict adherence to chronology might have been a logical approach, but would have caused the sequence to begin with Trist's 'flood diary', which is not the most accessible of the documents presented. Instead, the engaging, personal account of the floods that Trist wrote from his notes in his diary is presented first, followed by the history of flooding in Suffolk that he researched and wrote in tandem. These two chapters exist as a single typescript document in the archive. Extracts from his diary covering the extraordinary days of early February 1953 are then presented before the third and final text: the preparatory notes for the contribution that Trist made to the Ministry of Agriculture report on the effects of the sea floods on agriculture. These were largely completed in 1953, though some additions and corrections may have been made as late as July 1957. At the end of the

after the flood and published in 1959. Trist and his executive assistant, J. Simmons, contributed copy about agricultural operations (MAF 157/23, County Agricultural Relief Committee: Correspondence with Lord Mayor's National Flood and Tempest Distress Fund).

36 'The sea floods 1953 in Suffolk – diary of observations', below, p. 69.

37 'Big Fire At Melton: AEC Headquarters Gutted', *EADT*, Thursday 29 January 1952.

38 'The great sea floods of 1953', below, p. 3.

volume, 16 maps are presented for ease of reference and in order not to break up the sequence and flow of the preceding textual documents. Although completed in the early months of 1954, these annotated maps draw directly from observations Trist made during an aerial survey of the coastline on the afternoon of Wednesday 4 February 1953, which is described in detail in his personal account of the disaster.

Suffolk Archives, A2727/1/1. 'The great sea floods of 1953' and 'Records in the history of the Suffolk coastline'. A typescript draft monograph by P.J.O. Trist

This document is unbound. It comprises 52 sheets of thin paper, typed on both sides, with double-line spacing and hand-written corrections in ink and pencil. The cover sheet is dated 23 July 1957 and is signed by Trist. It comprises two chapters largely written in 1953. The first chapter, entitled 'The great sea floods of 1953' is a personal account of Trist's involvement in flood-monitoring and sea defence repair operations from January to August 1953, including an aerial survey of the coastline. The second chapter, 'Records in the history of the Suffolk coastline', is concerned with the long history of flooding along the Suffolk coast. It is an examination of historic efforts to improve sea defences and marsh drainage for commercial benefit, and the thorny issue of who was responsible for the upkeep of such. Written in tandem with the 1953 account, it was informed by Trist's personal research using sixteenth-century statutes, minute books relating to the work of the Commissioners of Sewers, and farm accounts found in the local archives.

Each page of the typescript is numbered in the top corner, and this has been retained within the transcription, in square brackets like this [*Page* 69]. Throughout the document there are hand-written amendments to spelling and grammar written in pencil and blue ink. These appear to be in the hand of the author and have been incorporated into the transcription largely without comment, but changes to the names of people and places are indicated with a footnote. In one place, a large section of the original text was obscured by a slip of paper securely stapled on both sides. The typed text on this overlay has been incorporated but the covered text has not been transcribed. With this approach to corrections, it is hoped that Trist's narrative account can be read most easily, in the form he had taken the trouble to create.

Suffolk Archives, A2727/1/2. 'The sea floods 1953 in Suffolk'. A diary of observations by P.J.O. Trist, 30 January–2 March 1953

This document is hand-written in a directory with sections for each letter of the alphabet, like an address book. Supplied by the Stationery Office (HMSO), the stitched and bound book with hard board covers, measures about 8 by 13 inches. Inside, the lined paper has yellowed, but the writing, in ink, is legible. However, it is very difficult to read in places because the handwriting is less than neat. Words are also heavily abbreviated at times, giving the impression that the author was hurried. There are many crossings out, annotations with arrows and asterisks, and inconsistencies in spellings and date formats. This all suggests that this was a working document set down in haste during the floods and later referred to by Trist when he composed his more ordered personal account of the 'The great sea floods of 1953'.

A written index on the inside cover indicates how Trist organised his diary entries and other notes in the alphabetical sections. The transcription in this volume begins with Trist's diary from section D (and continued in section G). Notes from sections K, C, F, E, and L then follow. Not all sections have been transcribed because where a subject is covered more fully elsewhere in Trist's papers (e.g. the dimensions of breaches in sea walls comprehensively plotted on maps), or the content is less interesting (e.g. telephone number lists), it has been omitted. For ease of comprehension, some content is presented in the order indicated by notations in the document, and not necessarily in the same sequence in which it was originally set down. For example, where an additional entry for 6 February was jotted in the middle of the page about 9 February, it is presented in this volume with the entry for 6 February. Where this has occurred, an explanatory footnote is provided.

Suffolk Archives, A2727/1/3. 'Sea floods 1953 – Report'

This document is an HMSO soft-covered, ruled, notebook containing hand-written notes in a mixture of inks. It is about 7 by 9½ inches in size. Production codes on the front cover date the stationery to 1950. Text on the front cover reads 'Sea Floods 1953 – Report' and 'Notes for the compilation of 1. Questionnaire for E. Suffolk'. The content relates to the surveying of flooded areas, assistance given to the river boards, methods of disposal of dead animals, recovery of farm equipment, and the organisation of men for the repair of sea defences. The way in which the notes are organized and annotated with chapter or section headings, suggest that this was draft content for the NAAS report, 'The Effect on Agriculture of the East Coast Floods, 1953', prepared by a working party set up by the main advisory committee on sea-flooded land. Trist served on this committee, and it is likely that he made or added to these notes for some time after the flood, as reference is made to 1955 events.

Suffolk Archives, A2727/2. Maps of Suffolk coastal lands

This is a collection of 16 photostat copies of Ordnance Survey (OS) maps, at a scale of 2½ inches to 1 mile. These maps show the extent of land inundated by the floods of 31 January and 1 February 1953, and include annotations relating to breaches in sea defences along the estuarine rivers and coastline of Suffolk. These were based on aerial observations that Trist made during a flight over the flooded areas on 4 February 1953, with further supplementary detail drawn from surveys conducted at ground level by his team. These maps are a striking, visual record that Trist completed in the early months of 1954. Each map is signed and dated.

Many of the places mentioned in Trist's written accounts can be identified on these maps and, where relevant, cross-references have been included in the transcriptions. To make a printable collection, uniform in size and scale, minor edits to the layout of some maps have been made. These edits include relocating the key, for example, and in some instances cropping off expanses of sea or land unaffected by the flood. The colour balance has also been enhanced to improve legibility. Great care has been taken to ensure no material changes were made to the mapped information.

Editorial method

Capitalisation (or lack of it) by Trist has been respected throughout. Original spellings have been retained, for example 'liason' and 'succour'. If text crossed out in the document remains legible, it is presented as strike-through text, ~~like this~~, in the volume. The exceptions are hand-written minor corrections to spelling and grammar, where it is clear the author intended the amendments to be made before publication. These have been incorporated without comment. Italicised text within square brackets [*like this*] is an editorial insertion added for clarity of meaning. Obvious abbreviations have been extended at times without comment. Footnotes in the original documents have been retained as footnotes in the transcriptions and are indicated by [*Trist's footnote reads*] to distinguish them from editorial comments. Pre-decimal money appears as £ s. and d., pounds, shillings and pence.

THE GREAT SEA FLOODS OF 1953

[*Page* 69]

~~Chapter 4~~ [Chapter] ~~5~~

"The great sea floods of 1953"

The story of the great sea flood as it affected the Suffolk Coast is long: although comparatively short in duration, it left tragedy, destruction and damage in its wake, which will leave its mark for several years to come. In writing a full account, I find myself a little handicapped, for my concern and responsibility at the time was directed towards the land and I had little first hand knowledge of what went on in the flooded coastal towns.

This record will therefore be largely concerned with the countryside events of which I had immediate experience; but first let me pay tribute to the heroism of many greatly fatigued men and women in all walks of life who tirelessly assisted in the rescue of those who were trapped and marooned. The deeds of some have been recorded; the deeds of many will be remembered, and of others will never be told.

During the evening of Friday, January 30th, a fresh strong wind was blowing in the north with a tendency to north east and it was strong enough to make the rivermen see to their moorings on the Deben. By the morning of the 31st, the wind had swung round to north west and was strong. At noon, I took myself to the Bull on Market Hill, Woodbridge and the wind was blowing hard straight through the front door.[1] Over big and small pots, everyone was discussing the gale and the foreboding of its quarter. Those who lived near the coast or rivers were guessing the possibilities if the wind should continue, for it was a period of peak spring tides at the full moon.

After lunch I had planned to carry a gun over Cross farm for a wild [*Page* 70] high wind appealed to me as an afternoon for shooting over some turnips, in which the pigeons were particularly interested.[2] I found my hide flat! In the teeth of a roaring gale, I tried to rebuild it but it was impossible to make a job of cover for as fast as a layer of bracken was woven into the network of large branches, it was blown away. The wind blew in terrific gusts and within a quarter of an hour, the hide collapsed on top of me! I roughly rebuilt it, but the gust came again scattering the hide along the hedge and I gave up the fight.

From about 4 o'clock the wind grew stronger. It blew in occasional gusts of terrific force which seemed to get stronger as evening approached. Up to about 11 o'clock when I retired to bed, the gusts increased in intensity like jet fighters screaming to earth in death dives. The windows rattled and the walls of the house shuddered spasmodically.

Between midnight and one o'clock as the moon and the tide approached their zenith, the wind force increased and the sea surge rose, until the river and sea defences could no longer hold back the waters.

I turned in to sleep and like many others out of harm's way from the flood, although living within a quarter of a mile of the Deben, slept like a log through a night when thousands of people on the coast and along the river estuaries were alerted to saving themselves or others.[3]

---

1 The Bull Hotel, Market Hill, Woodbridge, its landlord Douglas Watts, 1937–57.

2 Cross Farm, Waldringfield.

3 Trist was living at Dairy Farmhouse, Fishpond Road, Waldringfield.

On the morning of Sunday, February 1st, I roused myself earlier than usual and made for the window which overlooks the river. There it was – the worst had happened. River walls were broken, but I still had no idea that the damage had reached such proportions as our story will tell.

My first impression was the proximity of the river. Waller's marsh was full of water and the edge of the tide was in the garden of Novacastria about 250 yards away.[4] I dressed and breakfasted in a hurry and made off towards the river. On the saltings below the site of the old manor house at Waldringfield, [*Page* 71] the first thing which impressed me was the height to which the tide had reached. There was the debris as proof showing a tide mark 6 foot above normal. At this point of the river it had therefore been flowing 2 foot above the top of the river wall.

Waller's marsh lay in 8 feet of salt water with no hope of release by the little sluice until the tide had gone down.[5] I walked over the arable land of Cross farm where the tide had left and the ground was covered with thousands of dead worms and several moles. On the ground and in the air there was a continuous shrill scream of excitement as the gulls harvested an easy meal. Down in the village at the riverside there was a muddle. The ground floor of some of the houses were awash, whilst Mr. Nunn's boatyard had been lifted into confusion by the tide which swept over the quay.[6]

I looked over the river to the opposite bank and there saw disaster which particularly concerned me. The walls defending Cliff farm, Sutton and Pettistree Hall marshes were breached in five places and the land lay full of water. I remembered seeing this land flooded with salt water only a few years previously in the high tides of March 1st, 1949 and my sympathy again went out to the unfortunate occupiers.[7] At 11.30 a.m. the height of the tide was 2 foot below the top of the river wall with a little over two hours to go before the afternoon high tide level. This morning's level would be at least 2 feet above the average high tide mark. The wind had now abated considerably, but it still stayed in the north west danger corner. I watched the approach of high tide – but little did I know at that time that it would be difficult for much more damage to be done! An unusual phenomenon on this occasion was that the tide never left the river all day, and its height was about the same in mid-morning as it had been in the afternoon, on the Deben.

[*Page* 72] I returned home and through the medium of that uncertain friend, the telephone, I soon had a rough idea of what had happened to the land along the river estuaries and the coast. Meanwhile those who were not immediately affected had

4 Novacastria was a neighbouring house in Sandy Lane, built by Thomas N. Waller when he returned to the village of his birth on retirement from Hawthorn Leslie Shipyard in Newcastle, whose inhabitants are known as Novocastrians (Waldringfield History Group, *Waldringfield* (Chelmsford, 2020), p. 170). Both Dairy Farm and Novacastria are indicated on Map A2727/2/4, Kirton–Woodbridge, close to reference mark 22.

5 Sluice: a mechanical appliance, typically a sliding gate or similar, for regulating the flow of water in and out of a channel, for draining or irrigating land.

6 Nunn's boatyard on Waldringfield Quay (on the site of the old cement works) was established in 1921 by Harry Nunn. It was run in 1953 by his brother Ernest Arthur Nunn and renowned for building the first Kestrel sailing dinghy before the popularity of the class spread beyond the Deben to waters and boatyards across the country (*Waldringfield,* 2020, p. 266).

7 A major North Sea storm surge on 31 March 1949 led to what was described at the time as the worst coastal flooding in 65 years. Subsequently the 1949 event was characterised as the forewarning to the 1953 disaster that was ignored (D. Summers *The East Coast Floods* (Newton Abbot, 1978); P.J. Baxter, 'The East Coast Big Flood, 31 January–1 February 1953: A Summary of the Human Disaster', *Philosophical Transactions of The Royal Society A* No. 363 (2005)).

no idea of the hours of terror which many unfortunate people went through as they fled before the surge, or clambered on to their rooftops.

At Felixstowe, Ipswich, Aldeburgh, Southwold, Lowestoft and Yarmouth, the night was long! Not everyone concerned was aware of the approaching danger and some were caught in their beds. This account of the flood, is mainly concerned with the land, but I will describe some of the scenes of devastation at the coastal towns later in this chapter.

Monday morning came: and the next fortnight was to be memorable for many in whatever capacity they served to assist. Before telling you of the attack on the problem facing the land, let me first mention for the benefit of my readers who are not concerned with the farming world, some of the various responsibilities involved – which incidentally were not all completely cut and dried – as it is probably at least two centuries since this country last experienced a flood of such proportions – in fact, since Dunwich received its death blow as a city.[8]

The maintenance and repair of river and seas walls is the responsibility in Suffolk of the newly formed East Suffolk and Norfolk River Board.[9] The flood operation was therefore primarily their concern, but the disaster was of such proportions involving the employment of engineers, manual labour and equipment far outside their resources, that the Ministry of Agriculture stepped in to assist.[10] In Suffolk, the county was divided for operational purposes. From the Orwell to the Alde, the headquarters was at Ipswich with Mr. K.C. Noble, an engineer, in administrative charge as the Ministry of Agriculture Liaison Officer.[11] For the first few days, the Chief Engineer in charge was Mr. C.H. Dobbie, a consulting engineer, who a few months previously had been [*Page* 73] on the scene of operations at Lynmouth flood disaster.[12] During the course of the first week, several staff alterations were made as the operation settled into a pattern. More engineers were drafted in from the staff of River Boards all over the country. Mr. Dobbie was moved north and his place was taken by Mr. F.M. Andrews, of the Thames Conservancy Board, on February 8th.

I have vivid recollections of Andrews after his first day on the Deben. He returned late to the Bull Hotel at Woodbridge with his overcoat richly daubed with Deben's clay after trudging miles along the wall from Bawdsey to Ramsholt over

8 Often characterised as a lost city, in 1200 Dunwich, Suffolk was one of medieval England's wealthiest ports with a population of 5,000. A succession of sea floods in the late thirteenth and early fourteenth centuries drastically reduced its size and importance. As the river at Dunwich shifted its exit 2.5 miles (4 kilometres) northwards, sea defences were not maintained, and the town shrank to the small village it is today.

9 Under the 1948 River Boards Act the old catchment boards were replaced. Trist later writes that the East Suffolk and Norfolk River Board was not established until 1952.

10 Trist writes more about the operational basis for this. See 'Assistance to and liason with River Boards: staff and equipment' in 'Sea floods 1953 – Report', below, p. 102.

11 K.C. Noble, based at Upton Lodge, Warwick Road, Ipswich, sent daily 'Flood Reports' from East Suffolk and Norfolk River Board to the MAF in London, collating information supplied to him by field staff. Some examples are in MAF 157/25, Reports and surveys on extent of land flooded in East Suffolk.

12 On 15 August 1952 the village of Lynmouth on the north coast of Devon was devastated by a flash-flood caused by heavy rainfall. The event claimed 34 lives. The civil engineer C.H. Dobbie contributed to research and analysis of the disaster. In the 58th report of Scientific Memoranda of the Devonshire Association that he prepared, he included an estimation that the precipitation in that one day was sufficient to meet the water needs of Lynmouth's total population for 108 years (Lynton and Lynmouth's visitors website, '1952 Lynmouth Flood Disaster', 2022) The Lynmouth flood triggered a significant national response and discussions about the adequacy of flood prevention measures and infrastructure across the country, just five months before the sea floods in 1953.

mud squelching sand bags. On that side of the river he had made a survey of the position and felt happy that at least one little piece of the area for which he was responsible, was now "under his hat". Later in the evening he learnt of the possibility of his transfer to Norfolk and in his calm and placid way he remarked on reflection of his day's work – "After all that …!" And like a good servant stood by for further instructions.

Other engineers assisting Andrews were Mr. Adams (Chief Engineer, Isle of Wight) who took over the left bank of the Deben, and Mr. R.I. Lakin of the Yorkshire Ouse River Board who was on the right bank.

In the southern area, a sub station was set up at the Sudbourne Estate Office where Mr. M. Nixon (Chief engineer of the River Trent Board) was engineer in charge with Mr. D.R. Hindley of the Yorkshire Ouse Board assisting: their area comprised the Alde and Ore and the coastline down to Bawdsey. At Ipswich Mr. Whall and Mr. Miller of the county river board staff were assisting Mr. Noble with supplies and machinery, whilst Capt. Girton the Military liaison officer sat and calmly deployed the supplies and [*Page* 74] movement of the military force serving as emergency labour on the river walls.

The work in the northern half of the county was directed from the River Board's main office in Norwich. Mr. W. Roberts one of the county engineers was based on Aldeburgh, with Mr. H. Burton of the Mersey River Board, Mr. King ~~of the county staff~~ and Mr. Hollingsworth of the Thames Conservancy Board superintending the work from the Alde to Lowestoft. At the fountain head was Mr. K.E. Cotton, M.B.E., the Chief Engineer to the Norfolk and Suffolk River Board with the deputy engineer Mr. A.G. Alsop.[13] In the office sat two important 'back-room boys', the Clerk and Solicitor to the Board Mr. S. Vincent Ellis and his deputy Mr. C.E. Upson, who must have looked on in amazement at the mounting bill of accounts and wondered why the Board could not have armed itself previously with such audacity of spending!

As a first move the County Agricultural Executive Committee placed all its resources at the disposal of the River Board: its small remaining labour force and all the tools and equipment which were of service. The first job in the field was to assess the position, and members of staff were each allotted a sector of river bank or coast to survey. Reports came in by 'phone all day and the position grew blacker as one lot of marshes after another were reported as drowned. Late in the afternoon of this first Monday I totted a rough assessment of the situation and got a figure of 20,000 acres under flood.

Tuesday was a memorable morning dedicated to the god of communications – the telephone. Farmers in distress and in a chaotic muddle – 'had we got pumps, boats and men?' – the River Board asking for everything we could provide. After lunch I decided that the best thing to do was to get out, not only for the good of my ear and patience – but to see the position for myself. I started a foot survey of the right bank of the Orwell with my district officer, Harry Lea.

[*Page* 75] We started above H.M.S. Ganges at Shotley where Admiralty men were already at work on breaches.[14] We then moved on to the south of Cranes

13 K.E. Cotton presented a general description of the work done by the river board in Suffolk, including the standards adopted for reconstruction of clay walls in *Conference on the North Sea Floods of 31 January/1 February, 1953: a Collection of Papers Presented at the Institution in December 1953* (London,1954), pp. 200–11.

14 HMS *Ganges* was a shore-based Royal Navy training facility for boys located at Shotley Gate on the tip of the Shotley peninsula. It was operational from 1906 to1976.

Hill where there was a small but deep break south of the saltings. Between there and Collimar Point there was a fair bit of damage to the top of the wall and much inside scouring all along the wall to ~~Jill's Hole~~ Hares Creek.[15] The Strand marshes at Wherstead were flooded and there were two breaks in the wall.[16]

This gave me a slight picture – but compared with the damage yet to be seen, this side of the Orwell got off lightly. This took the best part of an afternoon, and I began to think in terms of a motor launch to speed up the survey, but early in the evening another idea was put into my head.

Mr. Alfred Adams of Felixstowe 'phoned me and although he and his brother Charles had several hundred acres of marsh under flood, he had time to think how others could be helped and suggested a friend of his who would place a plane at my disposal.[17]

At 2 p.m. on the afternoon of February 4th I drove up to the Ipswich airport at Nacton accompanied by Walter Hayles, where we met Mr. R. Jackson who had kindly offered to take us up in his Auster to get an aerial view of the floods.[18] A fairly strong N.W. wind was still blowing and of all the dull grey blustery days, this was a good choice for an introduction to flying!

We took off and headed towards to the Stour estuary and as we climbed to 700 feet the air was bumpy. My first recollections were that it is one thing to feel airborne and something more alarming to suddenly drop in space! Just how many times my sweet changed position with my meat course, I don't know – but I retained both! I have no head for heights and for ten minutes I only dared look ahead. There was only one thing to do [*Page* 74] and that was to concentrate on the purpose for which I had undertake this escapade.

With a notebook on my knee and Walter acting as an assistant to keep our bearings on a map, I started to record the damage. From the air, we had a wonderful picture of the devastation and of the early attempts of a small army of men feverishly active on the walls as the high tide rose on their temporary defences. There were men digging and filling sand bags, carrying timbers and tools whilst the tide surrounded them as they walked the walls now left as long thin broken strands in a great sea.

The damage on the left bank of the Stour was not great, but as we came out of the river I looked towards Essex and saw Harwich in great distress.[19] We turned inland up the Orwell and there lay the Landguard marshes inundated, with the top of the wall to the west of the old fort completely washed off, and a huge break just below

15 Scouring: wave reflection from sea and river walls may result in hydrodynamic scour; the removal of sediment such as silt, sand and gravel from around the base of the structure impeding the flow of water. Scour caused by fast-flowing water can carve out scour holes, compromising the integrity of a wall. Trist notes 'inside scouring' which is evidence water flowed over the top of the wall, to damage the side of the wall facing inland.

16 Map A2727/2/2, Ipswich.

17 Alfred V. Adams (1904–83) of Laurel View and Charles T. Adams (1901–77) of Roydon House, Marsh Lane, residences associated with the farm business, P. Adams & Sons Ltd (Laurel Stud), off Ferry Road, Felixstowe, and also at Red House Farm, Bucklesham. Alfred was a special constable (BBC website, '1953 Floods Eye Witness Accounts', www.bbc.co.uk/suffolk/dont_miss/floods/eye_witness_accounts/bernard_adams_fx.shtml).

18 Walter A. Hayles, a NAAS colleague, machinery adviser for Suffolk from 1946–70. In his diary Trist recorded the name of the light aircraft pilot as Mr W. Jackson.

19 Map A2727/2/1, Brantham–Harkstead; Map A2727/2/3, Erwarton–Felixstowe (Harwich is obscured by the key).

the hill separating the level from the Trimley marshes.[20] From the air, one could visualize the colossal oncoming surge of tide flowing in to the two rivers at this point with the nor' west wind meeting the flood after it had broken into the marsh. As I was later to see, the tide reached a height of 9 –10 feet as it swelled across the flat ground to create havoc and destruction in Felixstowe.[21]

Along the Trimley marsh wall, a small band of men worked at top speed. Inside the wall, the water lay calm and deep and looking inland towards the edge of the marsh, the flooded hedges of the lower upland gave an indication of the height of the water. Seagulls flew low, dived and fussed around the land where the waters had subsided. These scavengers of the sea were harvesting dead worms. This was a common sight all up the coast and they were in thousands exploiting a situation whilst food must have been difficult to find in a rough sea.

We turned into the Deben – the estuary which was battered worse than any [*Page* 75] part of Suffolk. We flew up the left bank to see a vast sea of over 2,000 acres stretching from Bawdsey to Ramsholt beyond Green Point. I tried feverishly to record the damage, but it was too great – it was not a case of counting of assessing the width of breaks, but a matter of estimating the total loss of the river wall for hundreds of yards.[22] Beyond the Ramsholt Arms, that charming pub in the pines and sand so well known to those who ply up and down the rivers, the little Keepers Cottage marsh lay in a flood whilst the ancient church stood sentinel over a situation which it had witnessed before. The Ramsholt Lodge marshes, those by Shottisham Creek, the Pettistree Hall, the Methersgate and the Sutton Hoo marshes were all under water.[23]

On the other side of the river, Woodbridge had mustered its men and was already well ahead with sealing the breaks in the marsh wall between Kyson Point and the Town.[24] The marsh below Martlesham Hall was full and as we went over the river opposite Cross farm, I had the impression that the old breach in the wall was a little larger. Over Waldringfield there was considerable activity around the shore bungalows and Nunn's boatyard. At Hemley the picture was much the same as we have known for some years.

The little estuary of Kirton Creek took a nasty knock and the water was lying back inland to Newbourn.[25] Below the Creek, all the marshes of Kirton, Falkenham and Felixstowe levels was a sea, and there was considerable damage to the river walls.[26]

Out of the Deben, we rose over Bawdsey cliff along the coast to Shingle Street.[27] All the coastal marsh was under water. The houses of the Shingle Town stood out on their island bank of shingle with the 600 acres of Hollesley and Oxley marshes in flood. We went over the mouth [*Page* 76] of the Ore at 600 feet, but we bumped badly and went up to 1000 feet. Every now and again, I asked our pilot Mr. Jackson, to go down lower. He was willing to do as we wished, but always added 'make the best of it whilst we are down, for you will not like it for long'! How right he was!

20 Level: a flat tract of land.

21 Forty-one died in Felixstowe. The inquest into 36 of the deaths was reported in *EADT*, 18 February 1953.

22 Map A2727/2/5, Felixstowe–Bawdsey.

23 Map A2727/2/6, Ramsholt–Butley; Map A2727/2/4, Kirton–Woodbridge.

24 See the large, hatched area of blue on Map A2727/2/4, to the south of Woodbridge.

25 See extent of inland flooding from the break in defences at reference mark 18, Map A2727/2/4.

26 Map A2727/2/4; Map A2727/2/5, Felixstowe–Bawdsey.

27 Map A2727/2/6, Ramsholt–Butley.

On this particular day, it was not comfortable at low levels, but in the brief period whilst at 500 feet I did get a good view of the damage.

Going up the Ore at 1000 feet at a modest 50 m.p.h. there was an amazing view in spite of relatively poor visibility. But it was a horrid sight. In front of us lay the Boyton, Stonebridge, Butley and the Gedgrave level, a total of 1400 acres inundated.[28] On the left side of Butley creek, the wall was smashed for 600 yards. Havergate Island was full and its walls in a bad stated. On Orford Ness, that peculiar shingle bank running down from Aldeburgh to the mouth of the River Ore, the Ministry of Supply buildings were up to their eaves in flood and there was no sign of life.[29]

From Orford to Iken, the flooded level of about 2400 acres was a pathetic sight and I remember comparing the scene with the Felixstowe level. In both areas over the past four years, the farmers had been breaking up marshes and tackling the drainage problem.[30] This land was coming into full production – and now after all the energy, enthusiasm and expense, it lay flooded in a medium which would poison its bowels for some years.[31]

On either side of the Alde the devastation of the marshes was complete.[32] There lay the marsh of Dunningworth Hall under salt once again. The Hazelwood and Aldeburgh Hall and Town marshes were all drowned. Between Aldeburgh and Slaughden we could see a muddle and a feverish activity of men.

Leaving the town, we saw that the marshes below Thorpeness and the area of fen around the Mere were all victims of the great wash which surged over the shingle. Over Sizewell we approached Minsmere, unmistakable with its old and new landmarks, the wind pumps.[33] Poor Minsmere! Once again she lay [*Page* 77] drowned in a sea of salt, her fate of only a few years ago. We flew slowly in a circle around the whole level and below us in the centre I saw the dead body of an area of marsh which only the previous year I had restored to health after its wartime salt flooding. To have said it was disappointing or unfortunate would hardly have described how I felt; it was a child which I had nurtured with much patience and care – and it lay dead below me.

One breach was in the sea wall whilst the main damage was done to the wall below the Minsmere cliff, where the Board were at work with repairs. This wall,

28 Map A2727/2/6; Map A2727/2/8, Gedgrave–Orford.

29 Orford Ness is the largest vegetated shingle spit in Europe and stretches for about 10 miles (16 kilometres) with a maximum height above sea level of around 13 feet (4 metres). There had been an experimental military presence on Orford Ness for decades (in 1935, a small radar team turned technical theory into a practical air defence system, before relocating to Bawdsey Manor). However, it was feared that the flooding of the Lantern marshes could lead the Ministry of Supply to abandon the site with the loss of up to 200 local jobs (*EADT*, 19 February 1953). In fact, between 1953 and 1971, Orford Ness was occupied by the Atomic Weapons Research Establishment, as UK nuclear deterrent forces became the cornerstone of Cold War defence policy. See W. Cocroft and M. Alexander, 'Atomic Weapons Research Establishment, Orford Ness', English Heritage Research Department Report 10 (2009).

30 A detailed discussion of the historic use of East Suffolk marshes and their reclamation is provided in Trist, *A Survey of the Agriculture of Suffolk* (London, 1971), pp. 103–40.

31 Salt from seawater flooding is absorbed by soil. Though much is carried away with drainage water some is retained, poisoning plant life and causing soil to become sticky or plastic. Farmers were advised to restrain from cultivating flooded land to mitigate the problems (A2727/1/8, MAF, Advisory Leaflet on Treatment of Land Flooded by Sea Water, March 1953).

32 Map A2727/2/9, Sudbourne–Aldringham.

33 Map A2727/2/10, Aldringham–Dunwich.

known to thousands who have had tea and bathed off the shingle beach below the sandy cliff, was topped with tamarisk originally planted by Mr. Rope's father, who preceded his son in the occupation of the Lower Abbey farm on the south side of the great ditch which runs through the level.[34]

The shingle beach was almost unrecognisable. Thousands of tons of shingle had been shifted by the tide over the wall into the marsh, but the tamarisk still held by its roots which had penetrated deep into the clay bed of the old wall.[35] As the sea surged down the line of the cliffs and turned the corner over the shingle, there must have been utter turmoil inside the marsh enclosed by the Coney Hill wall, which was badly smashed for almost 100 yards. The depth of reed debris which I later saw here and in other places, gave some indication of the might of the tide lashed into fury.

Over the cliff to Dunwich which could tell many a tale of bygone furies of tide and gale, we flew up the Dingle and Reedland marshes into the valley south of Walberswick to the Point and Westwood marshes.[36] This area of about 1,150 acres had not recovered from an inundation [*Page* 78] in 1949 and was also under salt water during the war: but a start had been made, for the old Dunwich river from Walberswick had been excavated beyond the Point marshes and the excavator was coming down the main ditch through the Dingle marsh.[37] There it stood with the sea water lapping into its cab.

The sky was overcast, it was very cold and visibility was poor just after 4 p.m. so we decided to return. We made height and got into a comfortable 'layer' of air. We had made a useful trip and recorded helpful information and this rather than curiosity was the purpose of the flight. We touched down at 4.30 and I have no hesitation in admitting that I was glad to get my feet on the ground.

For the first four or five days after the flood, it was a case of taking stock of the situation and obtaining equipment which had to come from all over the country. The task facing the responsible authority was tremendous. In the flooded coastal towns, the various civilian emergency services took over rescue and rehabilitation and did a wonderful job. Where the land was concerned, the great problem was of approach to the battered walls and this became literally a single file approach, temporarily sealing one breach and then on to the next. Military aid were the first to arrive in

---

34 The Ropes were a well-established farming family in Suffolk. It is likely that the tamarisk was planted by Arthur Mingay Rope (1850–1946), farmer and long-serving church warden. The St. Matthew window at St. Margaret's Church, Leiston, was dedicated to him by his daughter, the acclaimed stained-glass artist M.E. Aldrich Rope (1891–1988). Scenes related to the family's farming life on the marshes show ploughed fields in close proximity to the sea. In 1953 Geoffrey Austin Rope (1894–1976), NFU representative for Theberton parish, farmed Lower Abbey. The Rope family archive concerning farms at Leiston and Blaxhall is available at Suffolk Archives (HA412).

35 Tamarisk (*Tamarix*), also known as salt cedar, is a flowering shrub that is salt and drought resistant. Frequently planted as a hedge or windbreak in UK coastal areas, its long taproot and dense growth habit is useful for erosion control, particularly on dry, sloped areas.

36 Map 2727/2/11, Dunwich–Easton Bavents.

37 The wartime flooding at Dunwich was a defence measure. The long beach at Minsmere was considered vulnerable to a German landing, so the area behind the sand dunes was deliberately flooded to act as a barrier to any enemy who might try to make their way ashore. The area to the north of the Minsmere river was allowed to flood completely, while the area to the south was only to be flooded when an invasion was imminent. In June 1940 the sea sluice was opened and the door allowing the exit of freshwater from the river closed. When the whole of the north level was flooded, the sea sluice was closed and the area remained inundated until 1945, when the operation of the sluices returned to normal (Robert Liddiard and David Sims, *A Very Dangerous Locality: The Landscape of the Suffolk Sandlings in the Second World War* (Hatfield, 2018).

numbers and their approach presented a quick headache for billeting.[38] Before many civilians came on the scene, transport had to be arranged. There were excavators, bulldozers, dumpers, Duwks and other craft to be mustered.[39] The River Board staff of engineers was not adequate to cope with the situation and engineers were brought in from drainage authorities from far and wide.

Before we go any further, it would be as well to take stock of the situation briefly in order to give you some idea of what had happened and why there was such an urgent need to cope with the situation.

[*Page* 79] The following is a record of the acreage of grass and arable land flooded:

	Arable	Grass	Total	
River Stour	185	259	444	
River Orwell	356	1057	1413	
River Deben	1004	2576	3580	
Bawdsey–Shingle Street	112	623	735	
Rivers Ore and Alde	2192	3553	5745	
Aldeburgh–Walberswick	29	1585	1614	
River Blythe	146	1815	1961	
Southwold–Lowestoft	8	1049	1057	
~~River Waveney~~ Breydon Water	2	3958	3960	
	4034	16475	20509	Acres

Of the 4034 acres of flooded arable land under cultivation, 1934 acres was cropped with 1045 acres of wheat, 89 of oats, 100 of beans, 24 of kale, 486 temporary leys, 60 orchards and 130 of miscellaneous crops: the balance of 2100 acres was not planted. Almost every acre of these crops was destroyed or severely damaged and the longer the flood remained, the greater would be the penetration of the salt which would keep the land out of production all the longer.[40]

Apart from the damage already done both to the walls and the land, another period of high tides was due at the end of a fortnight after the flood.

The greater part of the damage was in the south of the county, more especially on the arable land. The northern area above Aldeburgh, although with a considerable area under water, had not suffered anything like the same damage to its sea defences. There were only three breaks in the wall on the Breydon Water, whilst much of the water poured over the top [*Page* 80] of the wall. There was one break on Oulton Dyke. The sea broke through the shingle at Benacre sluice, at Coverhithe and at

38 It was reported that, within a week, there were more than 2,000 military working on the walls (*EADT*, 11 February 1953).

39 DUKWs were six-wheel amphibious trucks, shaped like a boat. Capable of carrying 25 men or up to 5,000 lb of general cargo, they could maintain a speed of 5 knots (9 kilometres) in the water and 50 miles (80 kilometres) per hour on land. During the war, their primary purpose was to ferry ammunition, supplies, and equipment from supply ships in transport areas offshore to supply dumps and fighting units at the beach. They were also called ducks, which may explain variations in spelling.

40 Trist provides a summary of the duration of flooding in different areas of East Suffolk. See below, p. 32.

Easton Broad. The walls on the River Blyth and the shingle breaks at Walberswick, Minsmere and Slaughden were badly damaged. In the south below Aldeburgh, the walls of the Alde, Ore, Butley Creek, the Deben, Orwell and Stour suffered severe damage. On the Deben there were nearly a 100 breaks with a total length of just over one and half-miles of breaches.

This operation was one mad rush during all hours of daylight and for many, the organisation continued well on into the night, with a fresh start in the early hours of the morning. Each evening, the engineers, supported by the Army and R.A.F. liaison officers and members of the County Agricultural Committee staff assembled at the sign of The Bull in Woodbridge to consider plans and requirements. It was after these meetings when beer tempered the smell of salt, that I sat down to log what had happened and what I had seen during the day.

For the next ten days, the tempo of the operation increased daily until several thousand people were helping and a stream of equipment found its way to the coast. Let me continue from my diary where I recorded immediate impressions.[41]

February 5th. I spent a brief period in the office with several members of our staff transferred to this emergency. In my absence, the general order was to supply anything we had and do anything you are asked.

With a good layer of short warm jackets, I left for the Bawdsey side of the Deben. I looked in at Pettistree Hall and at Ramsholt and then pushed on to Bawdsey Ferry.[42] The Army and the R.A.F. boys were on the job in an endless file along the river wall. Standing on the bank, I looked across the marsh over one huge lake of salt which extended up to Ramsholt. Six [*Page* 81] feet of wind-ruffled water with an occasional top of a gate post standing out of the flood. It was a horrid sight.

Here was the general quandary: standing on the wall, on one side of me there was the river with its mud and saltings and on the other, the marshes full of water. An approach could only be made from either the Bawdsey or the Ramsholt end of the wall. The work had started and thousands of sand bags were being filled with mud and then piled up in lines in the centre of the wall. There was neither the time nor the labour available to make any attempt to fill the complete width of the breaks.

February 6th. I went down to Laurel farm, Felixstowe for the same purpose as my previous day's trip.[43] A first hand impression of the problems on the spot, would help in assessing what we could do to help and perhaps make some useful suggestions.

I found Mr. Charles Adams at the farm and he drove me down to the wall by the Golf Course. The wall up to the Ferry had been broken in three places and the Golf Course was flooded, but it was possible to drive a car through the water along the road crossing the course. We met his brother Alf and he showed me the position. Bags were being filled and carried two or three at a time along the wall to the breaks where there was water on either side. At this rate, it would be weeks before an area was sealed off. An army DUWK was floundering over the golf course trying to convey sandbags to the wall, but at each attempt its rear wheels became stuck on the side of a submerged ditch. It eventually succeeded in reaching its destination and a useful load of materials was landed.

41 See 'The sea floods 1953 in Suffolk – diary of observations', below, p. 67.

42 Map A2727/2/6, Ramsholt–Butley; Map A2727/2/5, Felixstowe–Bawdsey.

43 Map A2727/2/5, Felixstowe–Bawdsey.

Down at the Ferry there was a chaotic muddle, but order was being restored. The road was feet deep in shingle which had washed up from the shore bank and machinery was busy restoring the shingle to its place.

[*Page* 82] Large wooden huts had been shifted bodily for some distance, others lay on their side, whilst personal belongings were strewn everywhere. A gang of men were busy repairing the break in the wall which runs out at right angles to the river.

It was a bright sunny day but the wind was bitter from the north. We turned to the shelter of the Ferry Boat Inn. In the bar parlour all was restored to tidyness, but the walls bore a proud line at 4' 6", the height of the tide which had poured inside the house. It was past 1 o'clock and Alf had not yet had his breakfast; a halt for a few minutes would help, despite the inadvisability of Scotch on an empty tummy!

Holm Hill had once again been made an island and Mr. Brundish the farm foreman was marooned, as were Mr. and Mrs. King who were in the Commissioner's cottage * [*footnote in text*] at the Kings Fleet sluice.[44] They were taken off in boats which came over the flooded marsh from Laurel farm.[45] There were two farm horses standing up to their bellies in water, too frightened to move. Alf and David his nephew, rode out on two Suffolks and tried to encourage the horses to move inland – but they would not budge, so another attempt was made in a boat and eventually they were slowly chivvied out of the water.

These were not the only horses saved by Alf Adams for he was instrumental in saving the lives of nine others. At Kirton, after several men had given up hope of rescuing horses in a terrified condition and standing in deep water inside a barn, Alf rowed in alone and eventually encouraged the horses out of the barn, across the flooded marsh up on to dry land.[46]

February 7th. More troops had arrived and engineers were manning given sectors. With only eight days to go before the onset of further high tides, the position of filling the gaps was still serious. Mr. Noble, the Ministry's Liaison Officer phoned me to say he was now ready for all the labour we could find – and could [*Page* 83] the A.E.C. organise transport and spades and deal with any necessary catering arrangements.

During the afternoon I got in touch with the News Editor of the East Anglian Daily Times. 'Have you got any space?' I enquired. 'No!' came the answer. 'We never have enough!' But as soon as I told him of the urgency of an appeal for helpers on the river walls for the weekend, I was given the courtesy and help which this paper is always ready to give. The Ipswich "Green 'Un" was just 'going to bed', but the stop press was not full.[47]

---

44 [*Trist's footnote reads*] *This cottage was pulled down in 1953 to make way for the new large delph ditch. Delph, sometimes written 'delf', was a man-made channel, trench or drainage ditch behind an embankment such as a sea wall, on the landward side.

45 Brundish gave a brief account of his experience to the BBC programme, 'Focus on the Floods', broadcast on 12 February 1953. In it he explained that he and his wife were evacuated to live at Rues Cottages, Marsh Lane, Old Felixstowe (*EADT*, 13 February 1953).

46 Alf's son, Alfred Bernard Adams (1935–2018), recalled working with his cousin David to rescue horses at Kirton, without success (BBC website, '1953 Floods Eye Witness Accounts', www.bbc.co.uk/suffolk/dont_miss/floods/eye_witness_accounts/eye_witness_index.shtml, 2003). Perhaps the Adams were involved in multiple rescue attempts.

47 *The Green 'Un* was a supplement in the *Evening Star*, an Ipswich daily paper; sister title to the *EADT*, the county paper in circulation throughout Suffolk and north Essex.

Within half an hour of the paper being on sale, telephone calls started pouring in, before I had had an opportunity of warning Reg Bristow our Chief Labour Officer of the work that I was putting on his plate.[48] I drove over to see him and lay plans.

As I drove through the park at Boulge Hall, a thought turned me towards the churchyard where Edward FitzGerald lay and I recalled the peace of his lazy days up and down the Deben, when the world was contented – and returning to my mission, the calm vanished into a cold wind, with visions of mud and men with wheels and water everywhere.[49]

The plans were laid and the telephone rang – and rang, as the people from Ipswich volunteered service till midnight. I retired to bed for a brief period, when at 1.50 a.m. the police worried my phone. Someone was missing and there was a frantic wife. This was one of our men still manning the pumps on the wall at Bawdsey. I retired again till just before 7 a.m. when the phone rang to bid me the top of a bright and cold Sunday morning. At 8.30 I left for Pettistree Hall, Sutton where the volunteers were to assemble.[50]

Following the track which had already been badly cut into by constant [*Page* 84] traffic, I got to the wall to find 80 men at work. I imagine few of them had ever dug mud in such conditions in the face of a bitter wind, but they toiled willingly and did a magnificent job under the leadership of Mr. Warner the farm's foreman of the ~~Quilter~~ Bawdsey Estate, who by the time he had finished his job on the estate walls, had made no mean contribution in hard manual work and to the direction of hundreds of men.[51]

I left them to their job and with Walter Hayles, now with the honorary title of Master of the Boats (Deben) we drove off to Bawdsey to survey the position for launching boats on the flood inside the river wall so that food, stakes and sandbags could be ferried up to the men. The previous Saturday evening we had scoured the water front at Woodbridge for craft which had survived.

We returned to Melton and arranged transport to pick up a pontoon at Waldringfield. Here we were helped by Mr. Nunn who made light work of the job with his crane on the quayside. Early in the afternoon, the first of our fleet of 12 boats was launched into the Bawdsey marshes.[52]

Bawdsey on this Sunday afternoon presented a sight never seen before by Deben's side. Troops on the walls were battling with mud and bags. The Army engineers had laid a line and erected poles on the wall to carry electric lamps. Boats were ferrying all manner of loads and baulks of timber never designed for small craft. Of the three pumps which had arrived the previous day, one was working and the others were

48 R.A. Bristow, chief labour officer of AEC.

49 This is a rather poetic reflection from Trist as he recalls the Victorian era and a greatly admired former resident of Boulge Hall, the poet Edward FitzGerald (1829–83). Boulge Hall was demolished in 1955 when it became too expensive to maintain, but the poet's final resting place in Boulge churchyard remains.

50 Map A2727/2/6, Ramsholt–Butley.

51 The next morning it was reported that: 'East Suffolk people yesterday flocked in flood danger spots in the Bawdsey area, in response to Mr. Trist's appeal, published in the stop press of the "Ipswich Green 'Un" on Saturday night. Arriving by bus, car and bicycle the volunteers threw themselves into the task of repairing the deep breeches in the sea defences at Pettistree Hall Farm, Shottisham. One of the main tasks was filling sand-bags, and the work went on till nightfall' (*EADT*, 9 February 1953).

52 Details of the boats can be found in correspondence between W. Hayles and the East Suffolk and Norfolk River Board concerned with securing 'Official Orders' for the owners (MAF 157/25, Reports and surveys on extent of land flooded in East Suffolk).

being prepared. The business of manoeuvring heavy pumps along a narrow topped muddy wall was no mean task and had to be carried on incessantly throughout this operation. Below the wall were eight fire engine pumps and hundreds of yards of hose discharging water over the wall.

[*Page* 85] By late afternoon the temperature had fallen sharply with a bitter wind blowing fine snow. At dusk, a lone trail of men in single file could be seen plodding their way along the bank. They were service men tired, mud spattered and cold from a winter's day filling mud into bags.

Later in the evening I had news that the three breaks on the Breydon Water in the north of the county and the one south of Oulton dyke had been sealed.[53] Pumps were now in position and had a job which was to take five weeks.

February 9th. Late in the morning, I left for Martlesham aerodrome with Major R.C. Ridley the chairman of the A.E.C. and Sir Robert Gooch Bart., to meet the Minister of Agriculture Sir Thomas Dugdale, Bart., and Sir David Maxwell Fyfe the Home Secretary, who had been visiting the flood areas at Felixstowe.[54]

On Monday a further appeal for farm workers to help with the repairs, was made through the columns of the East Anglian Daily Times. The response from the farmers was immediate.

I went off to see how things were going in the Orford area.[55] At Boyton Hall there were 70 farm workers as well as troops and some men loaned by the Forestry Commission, who had got well ahead with filling in the centres of the eleven breaks and scours on the wall from Dock Farm to Boyton Hall: 145 workers were promised for the morrow. I went on to Gedgrave Hall, where the wall facing the Butley Creek had taken a nasty knock with serious damage to 596 yards, 200 of which was wrecked to saltings level.[56] By to-day about 200 yards was centre sandbagged. The position here was desperate. There was only one [*Page* 86] approach to the wall over 400 yards of a marsh mud track, after leaving the farm road and ploughing through nearly a mile of mud across the upland farm track. A hundred odd troops had been toiling hard but the gigantic task of sealing the huge gaps to stop the water coming in at each tide, was going to take a number of days.

I left Gedgrave at sunset. The marshes from Mr. Cordle's Chantry Farm at Orford to the south end of the Gedgrave Hall marshes facing the Lower Gull were full of water, 6 feet at the least.[57] As the sun fell, the fading light showed a deep calm on the face of the water; the tops of the hedges near the upland standing above like narrow banks, whilst the woods were gaunt islands. Beyond lay Orford Ness with the Kings and Lantern marshes filled with as much water as they could hold. A searchlight beam flickered over the water and the night drew on.

The Pettistree Hall wall breaks were completed today by local labour. The Master of the Boats continued the rallying of his fleet on the Deben.

53 Map A2727/2/15, Fritton–Burgh Castle; Map A2727/2/16, Hopton–Great Yarmouth.

54 It was reported that the object of the visit was to secure background information for a meeting of the emergency committee of ministers, to be held in Parliament that afternoon. The Felixstowe Dock and Langer Road areas were the extent of the ministerial tour (*EADT*, 10 February 1953); Major R.C. Ridley, was chairman of the AEC, 1948–54. He was succeeded by Sir Robert E.S. Gooch who took the chair until 1962. Sir Thomas Dugdale was Minister of Agriculture, 1951–54. Sir David Maxwell Fyfe was Home Secretary, 1951–54.

55 Map A2727/2/6, Ramsholt–Butley.

56 Map A2727/2/8, Gedgrave–Orford.

57 Samuel H. Cordle (1913–2003), tenant farmer at Chantry Farm, Orford.

It rained heavily overnight: apart from the intensely cold wind which had blown daily and the fine snow of the previous Sunday, this was the first weather to upset ground conditions.

February 10th. The East Anglian Daily Times gave prominence to an appeal for help. The 'Battle of the Breaches' was on – organisation had steadily gained momentum, the situation was now known and the various organisations were ready to cope in a big way. The column read – "Every available man is required to help repair sea and river walls. The position is desperate. We are fighting against time in face of the threat of approaching high tides .... I now appeal to all farmers in the county outside the flooded areas to spare whatever labour they can to help in this vital task!"

[*Page* 87] Whilst the paper was still hot from the Press, calls poured through to the A.E.C. offices at Melton where Reg Bristow continuously snatched the telephone until the connecting wires almost writhed in the heat of question and answer. We sent 55 men to Gedgrave's wall which no longer lay on its back, but had disintegrated for over a quarter of a mile into the water of the marsh: 34 went to Boyton, 17 to Shottisham Creek and 40 to the battered defences of Bawdsey, to join the R.A.F. boys and the army under the direction of Capt. Girton.

During the morning, the National Farmers' Union Executive had sat in Ipswich and stirred the broth of necessity for help. Later we shall hear of the response.

From Laurel Farm, Felixstowe to just north of the Kings Fleet, emergency repairs are within 2 foot of the top of the wall: 80% of the breaks in the Kirton – Falkenham area are in hand. The walls below Kirton Creek were being assaulted by chestnut pale fencing laid over the saltings.

This day, I went over the area between Orford and Aldeburgh. From the Boyton level I could see great activity on both sides of the Butley creek.[58] The support which the farm workers had given to the troops at Gedgrave was already making an impression. At Orford the break just south of the town was well in hand.[59] There were too few men on the Sudbourne bank, but access was difficult with a long walk along the wall.[60]

I met engineer Nixon returning from Poplar Farm, Iken where Mr. Piddington of the Admiralty was struggling with a handful of men to erect two pumps.

At Aldeburgh, the big break in the bulge of the marshes south of the town is temporarily abandoned for contractors to handle. In the office of the River Board at Aldeburgh I saw a plan of the depth of the scour [*Page* 88] which was 7 foot deep in the centre. Men were busy raising the old wall south west of the town to prevent the further inroads of the tide over the marsh.

The Ministry of Transport, the County Council, the local inhabitants and the River Boards are in one huge flap over the position at the Latimer Dam on the Lowestoft road.[61] The marshes are full of water on either side. The Benacre pumping station is hors-de combat and even it if is put right again in a few days, there are thousands of tons of shingle filling the final straight stretch of the Hundred river from which the pump operates. Yesterday a solitary excavator crawled along the sand dunes from the Kessingland Holiday Camp to make a token start on freeing the situation.

58 Map A2727/2/6, Ramsholt–Butley.
59 Map A2727/2/8, Gedgrave–Orford.
60 Map A2727/2/9, Sudbourne–Aldringham.
61 Map A2727/2/12, South Cove–Gisleham.

Towards evening Leslie Ripoll, who had accompanied me during the afternoon, drove back through the Iken–Sudbourne area.[62] The last light was flickering on the still waters of the sea which sprawled over hidden furrows and grass. By the road-side at Ferry farm, where the tide has receded from its furtherest bounds, the sand lay bleached and the extreme height of the tide was clearly marked by a meandering line of debris on the ground: and on the fence wires, hung sodden wads of reed. At Crag Farm, Sudbourne, a field of kale lay drowned almost to the top of the stalk. As I left Sudbourne smarting in its salt, I recalled how it was left after the war – after six years as a battle training area, with hedges high and ditches full of silt, with the land a mass of weeds and rabbits. Since 1949 the farmers, after an initial reclamation by the A.E.C., had restored this land to plenty – and now it was only fit for the Royal Navy as a base for landing barges. Such is the fate of lands beside the sea.

Later in the evening I joined the engineers at their nightly meeting at the sign of The Bull. This meeting and others were memorable. We sat deep in chairs with every edge and ledge supporting big and small pots of [*Page* 89] beer. The exchanges went on until the unofficial chairman called over from a standing position by the mantel-piece where he found he could rest his head against the wall to prevent it from nodding! This was K.C. Noble, the organiser of men and equipment who worked day and night with willingness, patience and understanding.

F.M. Andrews the chief engineer in the southern area of the county, informed us that from Bawdsey to Ramsholt the wall breaches were filled to within 2 feet of the top of the wall, but it was a thin infilling and he did not feel happy about the situation with the approaching high tides: the 2,500 acres north of Bawdsey were still full of water which Andrews estimated at 130 million gallons: enough to keep several pumps occupied for a fortnight at least, in addition to the exit via the sluices. At dark he had started his long trek to Bawdsey along the wall from Green Point after 1700 troops had plodded over the mud filled bags across the breaches. The traffic had been too great but there was no other way and the mud oozed out of the bags. Coming behind the men, Andrews sank over his knees in mud. He will remember the Bawdsey wall!

February 11th. Rain overnight: In the morning it was cold and overcast with a strong N.W. wind. Soon after 11 a.m. it rained in strong interval showers. Towards dark there was sleet followed by snow. 470 men from the farms turned out to fight the battle of the gaps. The real fight was on for the weather was making movement difficult and the low tides would soon mount up again. This was operation 'Boots, spades and grub'. All day, the receiver of calls at Melton would direct the labour to the banks with these significant words.[63]

Andrews phoned me at 10.30 a.m. The Bawdsey–Ramsholt line was fairly safe bar three breaches. These had suffered the previous evening when [*Page* 90] the Army trekked home in the pouring rain. Break No. 13 at Shottisham creek was 79 yards long and could only be reached by a 300 yard walk along the wall.

The Deben tide was 1' 3" higher than predicted today.

By 11 o'clock, Andrews joined me at Melton and I took him for an inspection of part of the right bank of the Deben and later over to the Orwell. At Kirton

---

62 Probably the colleague listed as C. Ripoll, assistant district officer in MAF memoranda (MAF 220/9, Correspondence between East Suffolk AEC and MAF Floods Emergency Division, Whitehall); Map A2727/2/9, Sudbourne–Aldringham.

63 This may have been R. Bristow the chief labour officer or other colleagues. There was a rota in place to ensure the phone was answered 24 hours a day, during the emergency.

Creek there were 4 breaks totalling 84 yards and 4 yards off the top of the wall by the sluice.[64] On the other side of the creek the top of the wall had been damaged above the concrete blocks. Here we met Engineer Lakin who had a problem with the sluice door.[65] At Kirton there was the most spectacular piece of emergency wall repairing done on the Deben. With the help and direction of local workers who were by no means new to this work, 400 boys from H.M.S. Ganges of Shotley had done magnificent work with the perfect combination of local skill and the trim finish of a job done by the Navy. They had not contented themselves with infilling the centre of the wall. Large lumps of concrete had been placed orderly at the base of the wall and the bags of clay were built up to the top for the entire width and depth of the wall. It was a pretty job and a good job.[66]

We spoke to some of the local men still at work and commended them for what they had done and compared it with other work which seemed 'very emergency'. 'Weem, a' doing a proper job', said one of the men – 'Do you jist fill the middle, the tide'll be in agin.'

The water was almost off the arable to the north of the creek, but the ditches were full. Large puddles lay on the land and in its sodden state, its reappearance from the depths had the ghastly look of a drowned face.

We went on to Levington Creek where about 40 acres lay drowned by the tide which poured in through a break of 29 yards.[67] From the centre of the bed of the [*Page* 91] wall to the back there was a scour varying from 6 to 10 feet in depth. At low tide, work could start if sandbags were prepared but other work involving large areas was still in danger and the work had to be deferred.

Over on the Orwell, the 5 breaks by H.M.S. Ganges had been sealed. The break below the oyster beds at Crane's Hill was three parts sealed with more good work from the Ganges boys. A small gang of men were working on the wall at Over Hall, Shotley, but there was no work going on between Hare's Creek and Collimar Pt. where the damage was largely to the top of the wall with back scouring.

From elsewhere I had news of the Adams' operations at Felixstowe. We had sent 100 men over to help and reports indicated that Alf had two tails today! Boats are operating north of the Kings Fleet. This is Walter's navy, comprising 3 pontoons, 3 motor boats, 2 whalers and 2 barges which were mustered from watermen on the Deben.

On the River Stour, an excavator is at work on the wall at Ness Hall, Erwarton, whilst Frank Keeble of Brantham Hall is quietly getting on with his own work.[68]

64 Map A2727/2/4, Kirton–Woodbridge; Map A2727/2/5, Felixstowe–Bawdsey.

65 R. Lakin of the Yorkshire Ouse River Board.

66 Trist was not the only person to appreciate the effort of the naval trainees that day. A dramatic photograph in the newspaper shows Admiral Sir Rhoderick McGrigor descending from a helicopter on a line, watched by boys on the beach. The First Sea Lord was lowered to the ground twice in the Kirton Creek area to inspect the work of Ganges boys and other helpers in repairing the sea wall below Red House Farm, Falkenham, and near Corporation Farm, Falkenham (*EADT*, 12 February 1953).

67 Map A2727/2/3, Erwarton–Felixstowe.

68 The Keebles had farmed in East Suffolk since the late nineteenth century. Francis Roland (Frank) Keeble (1895–1960) and his father John R. Keeble (1867–1946) developed one of the leading flocks of Suffolk sheep in the country, which were sent all over the world including Australia and America for stud purposes. John Keeble was also one of the four founders of the Suffolk branch of the NFU (*EADT*, 28 May 1946).

February 12th. More rain overnight. The morning was cold with intermittent rain and sleet and in the early afternoon there was fine snow which came down thicker as the temperature fell towards evening.

How the telephone rang to-day! Calls came in fast enough to detain five of the staff in my office to cope with the situation! Offers of men from farms all over East and West Suffolk poured through.

'How many do you want?'

'What should they bring?

'Where should the go?'

In the past three days, 1,837 farm workers have manned the walls.

[*Page* 92] The response for help has been magnificent. In the past, when kings required ships, Suffolk built them and manned them; on countless occasions the men of Suffolk have rallied to a cause and in this year of grace, her men came again – this time to the succour of her own folk.

At this time I made several notes of things I heard in praise of these workers and of the trouble which many farmers took in organising bus loads of workers from their districts. No help was scorned and each man played his part; and there was that obvious fellow feeling between the farm worker and the land which supported him and his family. Would he have been similarly useful and so full of resource in fighting a factory fire? Some were asked how the job was going, whilst they stood plastered in mud in the teeth of wind driven snow. "Do you try sugar beetin' uvver Stowmaar-ket when 'tis wet in Novemmer – than 'tis cauld".

I saw many hundreds of these men at work. One held a bag and another a spade. It cut into the mud; it dug down and was pushed over and pulled back to give a release from the suction. Here were men working in conditions which they understood from experiences in the ditch. How useful and how much easier it all is when you know exactly what you are doing!

What else was doing on this day? My diary records a first telephone call at 6.30 a.m. Labour ahoy! By 2.30 p.m. 1100 men were positioned for the morrow. The breaks on the north side of Martlesham creek had had no attention and during the morning a foreman of the River Board rang up in desperation for help. Leslie Ripoll went out to see the position and found a handful of men struggling on their own. The foreman reckoned he wanted 200 men – we arranged for 50 for the next day.

Inquiries from the River Blyth area in the northern sector indicated that they were going on fairly well but could do with help. We found them [*Page* 93] 83 men to report to the White Hart by Blythburgh Bridge.[69]

In the evening at the sign of The Bull inWoodbridge, the week-end arrangements were read over to the Engineers. 700 troops would be standing by on emergency call, if the approaching high tide should again overstep its mark. The Bawdsey R.A.F. men were to stand by in their own area. There was a special instruction for Mr. M. Nixon, the engineer in charge of the Iken-Orford sector: he was to be airborne at dawn on Sunday 15th by helicopter based on Sudbourne and was to fly the whole of the coastline down to Bawdsey, reporting by radio to the ground, and have troop dispositions know so that men could be called to emergency positions.

'Very good Sir', replied the amiable Nixon with tankard aloft, to the nodding head of Noble still smiling as it regained its prop against the wall. 'And the rest of you', said Noble, 'will patrol your own banks as from daylight on Sunday morning –

69 Map A2727/2/11, Dunwich–Easton Bavents.

and just one word of warning – you may get cut off on the wall if you are the wrong side of a new break!' The briefing was received with acclamation and expressions appropriate to such an occasion – followed by long deep draughts of beer.

February 13th. A Friday, but nothing untoward happened.[70] The 3 breaks which had caused the flooding on the Martlesham Hall marshes were sealed off to-day. The wall at Shottisham Creek was secured to the first stage of repair. It did look an odd line of repair: there had been serious back scouring damage with huge gaping holes behind and through the bed of the wall, so that the line had evolved into a series of temporary horseshoe bends as sandbags had to be laid out on the solid ground of the saltings. An excavator was busy at the back of the wall, floundering in mud as it grappled at the nearest soil to fill the giant cavities.

[*Page* 94] There were nearly 300 men working in appalling slush under the leadership of Mr.Warner of the Bawdsey Estate. The marshes were clear of flood water to-day.

On the Ramsholt Lodge Marsh, the flood had left, but the ditches were full. There had been 8 breaks and other damage to the top of the wall. The marsh lay strewn with huge lumps of clay as it had burst out of the wall.

At Bawdsey, the great assembly of cars, lorries, fire engines and troops had almost vanished. Two pumps on the wall were working hard and about half the height of the gate posts was now showing out of the water: about 2 foot 6 inches of water lower than last Sunday the 8th. The main force of labour today had concentrated on 150 yards of shattered wall below Poplar Farm.[71]

All through the week, the Ramsholt–Bawdsey wall had been a memorable sight at dusk. With the river to one hand and the flood to the other, the thin line of the top of the wall was entry and exit for everyone. As light fell, against a background of grey sky and snow-covered uplands there could be seen an endless file of men trudging with spades and mud-caked boots along the snow dappled wall – not in a straight line, but snakelike, as the hundreds of feet followed this meandering line of the wall. Surely the peace of Bawdsey's marsh had never suffered such a disturbance.

News from Sutton Hoo at the head of the river intimated that their breaks were sealed with the help of men we had sent to-day. Apart from local farm workers, the total labour force on the walls between the Alde and the Orwell [*Page* 95] today was 1168 farm workers from inland, 610 Army and 80 R.A.F. men.

In the evening I heard from Sir Peter Greenwell.[72] We had sent down 180 men to-day and they had done a fine job assisted by a small handful of troops. There were

---

70 Reports this day emphasised growing tensions about the new spring tides, and fears were accentuated by the looming prospect of a heavy snow fall. The news was that, 'as a Northerly Blizard swept East Anglia last night flooded areas faced the threat of the possible effects of a six-day period of high spring tides. Throughout the day winds of near gale force whipped the coastline. Rain turned to snow in most places as the temperature fell during the afternoon to almost freezing point.' The report warned that, 'Highest water levels will occur during the 24-hour period beginning at noon next Monday … peak danger times will be from about three hours before high tide to some two hours after' (*EADT*, 13 February 1953).

71 Map A2727/2/6, Ramsholt–Butley; Map A2727/2/5, Felixstowe–Bawdsey.

72 Sir Peter McClintock Greenwell (1914–78) farmed the Gedgrave estate acquired by his father in the 1930s. In the 1950s he was a county magistrate, becoming a prominent figure in local agriculture; he held the presidencies of the Suffolk Show and the East Suffolk branch of the CLA and was chairman of the Ministry of Agriculture's sugar beet research and education committee (*EADT*, 4 December 1978).

4 more breaks totalling 70 yards to seal. At the beginning of the week, the Gedgrave wall lay wrecked with 696 yards of wall in a ghastly mess.[73]

At the evening meeting, we roughly assessed the flood water still on the marshes: Aldeburgh 500 acres, Gedgrave 1,000, Iken 900, Boyton 100, Sudbourne 2,000, and Hollesley 200 acres. A total of 4,700 acres. Above the Alde, I assessed a further 7,000 acres still under water and this is the 13th day of the flood. And out of the mouths of engineers and other wise men, came forth two things worthy of record in this drama. Squadron Leader Cumber the O.C. at Bawdsey, who was a tower of strength in this operation, had this day caused to supply 2,800 pints of tea to his men! Each of 700 men, had drunk 4 pints in 4 hours: and in this catering which was conveyed by boats inside the wall he had cursed poor Andrews for allowing the pumps to go at full blast, thus creating a low water take off for the boats at Bawdsey! But this had not been the end of it for Andrews, he had also been cursed by Mr. Walker, the Estate Agent, for not getting the water pumped quick enough! What was a poor engineer to do?

A late report from Mr. Geoffrey Girling of Reydon Hall indicated that 200 men had been on the Blyth river walls to-day and all the work was in hand.[74] There had been 25–30 breaks or serious top scour damage.

February 14th. The tides have turned to rise again, but the weather is calm. There are still vast areas of flood, but all the breaks are sealed temporarily with the exception of one in the Aldeburgh Town marshes and one on Levington Creek.[75] If the wind remains calm, all will be well with the [*Page* 96] rising tides.

During the morning of telephoning and discussions on plans for the coming week, we started to sort out some details which in the rush of the past fortnight had been left to look after themselves. We had cleared ourselves of a vast stock of small tools and chartered buses, vans and lorries, right, left and centre; the day of reckoning must come, now is the time for orderly decisions to assist the backroom boys in their booking.

A telephone call from Lord Stradbroke conveyed the appreciation of H.R.H. the Duke of Edinburgh who had visited Bawdsey on Friday, for all the work which had been done.[76]

---

73 Map 2727/2/8, Gedgrave–Orford.

74 Geoffrey N. Girling (1898–1971), of Old Hall, Reydon, was NFU representative for the parish.

75 There was much concern about Aldeburgh on this day. It was reported that, 'not one half of the emergency wall being built across the marshes at Slaughden to prevent flooding of the town' had been completed due to the scale of the work required. After DUKWs arrived on site, it was reported that 'men and machines were working under arc-lights last night with their backs to the sea, for at this danger point only a 65-yard strip of land lies between the flooded marshes and the sea. Soldiers were knee-deep in water passing sandbags from hand to hand to build that part of the emergency wall near the pumping station' (*EADT*, 14 February 1953).

76 John Anthony Rous, 4th Earl of Stradbroke (1903–83), was Lord Lieutenant of Suffolk from 1948 to 1978. He demolished the family seat of Henham Hall, near Blythburgh in 1953 due to its costly upkeep. The Duke of Edinburgh flew into the airfield at Martlesham Heath, where he was greeted by the Lord Lieutenant and other local dignitries, then driven by car to see the work underway on the sea walls in the Bawdsey area. It was reported that His Royal Highness tramped at least 2 miles (3 kilometres) through thick mud and sand. In Sutton High Street he paused to gaze at the sand piled up to the first-floor level and commented: 'The job is far bigger than I ever imagined'. Press photographs taken during the royal tour of the Deben estuary show the Duke walking on snow-covered tarmac at Martlesham Heath airfield, talking with the Commanding Officer of Bawdsey RAF Station (Squadron Leader J.A. Theophilus), and inspecting the newly-made sandbag wall at Green Point, near Ramsholt (*EADT*, 14 February 1953).

After lunch, I made a 'round' upstream from Waldringfield, with a first call at Martlesham Hall.[77] The 3 breaks had been sealed and a few men were still working on the wall by the sluice. About a third of the marsh was still flooded. Below the wall, I saw a pile of concrete slabs, a wheel barrow and a spade standing in the flood: they were tools of trade, evidence of recent repairs to the creek wall before the great tide.

I drove on to Kyson Point. The marshes below Woodbridge were now free of water. The 4 breaks and 5 scours to the top of the wall had been promptly tackled by the men of the town immediately following the flood. It was low tide and at its edge, teal and wigeon fussed in the mud. They sprang and circled over the water, whilst the shelducks cackled and joined them in a leisurely flight: gulls were dipping over the water, until their formation was broken up by the clumsy flapping of a cormorant which took off in an upstream direction. As I left the wall I heard the whine of powerful wings and three swans passed over in line ahead and made towards the old tide mill.

Around the head of the river I turned into the sandy track which leads up to Sutton Hoo farm. A few days previously I had listened to a voice in [*Page* 97] desperation which said 'our little bit of marsh is in a forgotten corner of the river'.[78] We had come to the rescue and sent over 32 men and this afternoon the 3 breaks and 7 top scours had had good first aid repairs.

The sun was sinking as I walked across the sodden marsh to the river wall. Opposite, lay Woodbridge with its snow clad roof tops. From under the wall, I could hear many tongues chattering out of the mud and as I showed myself, the sentinel curlew whistled the alarm, and they got up. Teal sprang and the redshanks piped warning, whilst the shelducks just turned to look and waddled off at ease further along the mud.

The sun was now sinking in a great red ball over Kyson Point and I turned to face the hill of Sutton Hoo. Below me, stranded on the saltings, lay a derelict hulk and up on the hill lay the earthly burial ground of the great Viking ship which yielded rich treasure.[79] No doubt that ship and others which may be still buried on the hill, sailed up the Deben and over the marshes which are now enclosed.

Tonight the high tide is at 1 a.m. – a most inconvenient time for anything to happen. Emergency staffs of all the organisations concerned are at call – phones will be manned throughout the night – the teleprinters are kept warm – radio communication is set and the helicopter will hover at dawn. Approximately 7,000 acres are still under water.

February 15th. The third Sunday. It was calm. Snow was still on the ground and there was a further slight fall before noon. The Deben was full but the tide rose in an orderly manner. This was the 15th day of the operation and the first day on which a halt was declared. It was time to take a breath. My diary records that after listening to Country Magazine broadcast from the Sun Inn at Woodbridge – I fell asleep.[80]

[77] Map A2727/2/4, Kirton–Woodbridge.

[78] Trist writes more about this conversation in his diary. See p. 80 below.

[79] Sutton Hoo, where the discovery of a ship full of extraordinary treasures in 1939 revolutionised understanding of the Anglo-Saxon period.

[80] Trist writes more about this broadcast in his diary. See below, pp. 81–2.

[*Page* 98] The reader must bear with me in this chapter for the long story is difficult to treat. The disaster in Suffolk, covered an area of 20,500 acres and involved the whole length of the coast line. It was not possible to record the daily happenings everywhere, but I will attempt to build up a picture taken from first hand accounts of my own and others.

Let us first clear our consciences. This was not a calamity resulting from negligence.[81] There is no doubt that many lengths of tidal river walls were in need of strengthening and if some conditions had been better, the damage to the walls might not have been so extensive – the tidal assault was not against the outside of the walls, for it rose so high and was whipped to such fury by the gale, that it was flowing at least two foot over the top of the wall. The bulk of the damage done, was by wind and water raging inside the marshes, tearing at the weakness inside the walls.

If we start on the River Stour we can work north to examine what did happen. There was one break S.E. of Marsh Farm of about 30 yards and another of 15 yards south of Brantham Court and another of 10 yards to the east: one small break south of Queech Farm, Stutton and 2 below the Royal Hospital School, Holbrook.[82] There were 4 between a point S.W. of Beaumont Hall and Johnny All Alone, of about 60 yards in total length and 6 of about 60 yards between Ness farm and Creek House.[83] A total of about 190 yards of wall on the left bank of the Stour, flooding 185 acres of arable and 259 of grass marshes.

From Bloody Point on the right bank of the River Orwell, up to Ipswich there were 8 breaks and a dozen or more lengths of damage to the top of the wall: a total of about 350 yards of damage, was a rough estimate I made on inspection, several hundred yards from the walls but it is likely that the real [*Page* 99] length of damage exceeded 500 yards.[84] Four breaks near H.M.S. Ganges flooded some arable and the adjoining marsh of Over Hall, Shotley and the level up to Cranes Hill where there was one break of 35 yards just south of the old oyster beds. Up to Collimar Point, Jill's hole and round to Hare's creek, there was considerable damage to the top and the inside of the walls. Once again, the Hill House Farm marshes occupied by Mr. Donald Wrinch were drowned – their second salt baptism since 1949.[85] There were 2 breaks in the Strand marsh wall and here the water flowed over the road and on to Mr. Colwall's arable land on Wherstead Hall.[86] Along this bank of the river, 132 acres of arable and 307 of grass were flooded.

81 There was discussion at the time about who was to blame for inadequate river defences. Many river walls were in private ownership, and it was suggested that they had fallen into a state of disrepair due to the combined negligence of the authorities and land owners. As far as owners of marshes were concerned, they were quick to point out that they were under no obligation to maintain their walls as a public service, but only in so far as they wished to protect their own land. However, this was beside the point. As Trist explains, the water was so deep that the condition of walls was not a factor.

82 Map A2727/2/1, Brantham–Harkstead.

83 Map A2727/2/3, Erwarton–Felixstowe.

84 Ibid.

85 Donald James Wrinch (1885–1970) of Shotley Hall estate, also farmed Hill House and Red House farms, Shotley, formerly part of Woolverstone Hall estate, until 1937.

86 Map A2727/2/2, Ipswich; Wherstead Hall, formerly part of the Bourne estate was acquired by the Paul family (the local millers and grain merchants R. & W. Paul Ltd) in 1936. Philip Colwill (1878–1955) was a tenant farmer who is said to have relocated his entire farming operation (men, families, animals, riggings and all) from Bideford, Devon, to Ipswich, by train in the early twentieth century (History of Suffolk Food Hall website).

On the left bank of the Orwell, the toll of disaster was fairly heavy.[87] There was one break of 29 yards on Levington Creek. On the Trimley marshes there were 3 breaks between Fagbury Point and the Cliff of 38 yards: and between the cliff and the old fort by Walton, another 3 of 50 yards. Between Felixstowe Pier and Landguard Point the sea flooded over the top of the wall for 1870 yards and between Landguard Point and the coastguard station for 2433 yards, leaving only the high ground of Landguard common high and dry. It was in this area of the Languard marshes where the sea swept in to cause the terrible loss of life and property in Felixstowe. I had a look at these marshes after that water had left and the tide mark left on the thorn hedges was at 9 – 10 foot.[88]

The havoc and the mess in Felixstowe could only be seen to be appreciated. The surge, as it swept up Langer Road as far as the Ordnance Hotel, hauled pre-fab houses from their foundations and one was carried 200 yards from its site. Hundreds of houses were flooded through the first floor and many people were marooned for some hours. There were feats of rescue and duties devotedly fulfilled in Felixstowe which will never be written, but [*Page* 100] which will be remembered in gratitude.[89]

Of the flooded farm land, there was 224 acres of arable and 750 of marsh. Before leaving this river, you should hear something of flood borne debris, an expression which will crop up from time to time. On the Searsons marshes at Trimley occupied by Mr. Bernard Smith the raging tide dashed into the sides of a 'fleet' and tore whole pieces of marsh, complete with soil and rotting fibrous material on which reeds and coarse grass grew.[90] This material is the immature stage of peat formation and consists of a tangled mass of living reed and rush roots with decaying organic matter. Enormous chunks of this debris 2½–3 foot in depth, lay all over the marsh, one block measuring 200 by 55 yards. An estimate of the total quantity was 5,000 tons and this had been lifted by the tide, borne over the marsh and deposited several hundred yards away from its original site!

Following the coastline in front of Felixstowe, we turn into the River Deben.[91] I recall my first impression when I flew up the estuary two days after the high tide. The whole of Deben's marshes were awash and her walls were shattered. At Bawdsey, on both sides of the river, the marshes were two vast lakes as far as the eye could see.

The wall against the river at the Ferry had been breached, with 3 other breaks in the wall between the golf course and the Laurel Farm marshes of the Adams brothers. About 150 acres of winter corn which looked promising in November were drowned together with 91 sheep. There were 11 further breaks beyond Laurel Farm and another 11 in the inlet of the Kings fleet sluice: a further 10 or 12 in the

87 Map A2727/2/3, Erwarton–Felixstowe.

88 According to the harbour master at Harwich the tide reached a height in excess of 13 feet (4 metres) above mean sea level (*EADT*, 18 February 1953).

89 Lives lost at Felixstowe were publicly commemorated in 2006 with the installation of a memorial wall and garden. It is located near the junction of Langer Road and Beach Station Road, where the 1953 flood waters were at their highest. Some survivors' stories are recalled in Jean Macpherson, *The Felixstowe Floods of 1953: Never to be Forgotten* (privately published, 2023).

90 Trist describes fleets as 'the remains of old river beds, which are now wide ditches with boggy reed covered land on either side' (*A Survey of the Agriculture of Suffolk* (London, 1971)). Bernard C. Smith (1902–83), The Hall, Walton, was a member of the NFU County Executive Committee and local representative for the parishes of Trimley St. Martin and Trimley St. Mary.

91 Map A2727/2/5, Felixstowe–Bawdsey.

wall on the Falkenham level up to Corporation Farm. Here again there was a mess of debris torn out of 'fleets'. On the Kirton Creek, there were 4 breaks and two places of bad damage to the top of the wall.[92] There was also damage to the wall at Martlesham Hall and the Woodbridge marshes by Kyson Point which are previously [*Page* 101] recorded.

On the left bank, damage at Sutton Hoo and 4 breaks at Methersgate old quay. The walls against the Pettistree Hall marshes occupied by Mr. T. Miller had 9 breaks with a total length of about 410 yards.[93] From Cliff Farm, Sutton just above Pettistree Hall marshes to Bawdsey ferry, the walls were smashed to pieces. The walls of the Ramsholt Lodge marshes of Mr. Norman Simper by Shottisham Creek were breached in 5 places, and on the marsh below the Lodge, there were 8 breaks.[94] The wall below the Keepers cottage and Ramsholt church had 5 breaks and 7 places of damage to the top of the wall. From Pettistree Hall to Ramsholt there was 760 yards of damage. From Ramsholt to Bawdsey Ferry, the wall was shattered. Below Poplar farm 833 yards of wall was washed away. In this stretch, there were 69 breaks the longest being 200 yards and a total damaged length of 1211 yards. On the left bank, from Sutton to Bawdsey, the total length of damaged wall was 2077 yards.

Of the land flooded, there was 277 acres of arable and 1380 of grass, from Sutton Hoo to Bawdsey and from Felixstowe to Woodbridge, 727 acres of arable and 1196 of grass: a total of 3580 acres flooded on the Deben where there was also a relatively heavy loss in livestock of 120 cattle, 303 sheep, 18 pigs, 1114 poultry and 7 horses.[95]

As witness to the fury of the wind and tide on the Deben, the flood borne debris on the marshes between Middle Barn farm and the southern boundary of Ramsholt Lodge, was an amazing sight. Within 50 yards of the road opposite the entrance to Bawdsey Manor, there was a chunk of marsh which averaged 25 yards in width by 70 yards in length and a yard deep. A solid piece of reed and rush growth with a foot of clay on the bottom lay deposited on the marsh with the reeds unbroken in their original [*Page* 102] growing position. It had come from the Queens 'fleet' across three hundred yards of marsh! I have previously described these 'fleets' as the remains of old river beds, which are now wide ditches with boggy reed covered land on either side. The tide in its surge across the marsh must have dashed into the ditch and torn out pieces on the opposite bank as the pressure of water pushed forward. Enormous pieces of this debris lay over barb wire fences and across wide marsh ditches, so completely covering the span that one could walk over a ditch and not know of its existence. Below High House, Bawdsey, there is a marsh boundary of post and rail fencing standing 4 foot 6 inches high and at the back of the marsh, lies thousands of tons of this debris. It is probably an accurate guess to say that the tide must have been running 8 feet deep for this debris to have been carried over this fence.

Out of the Deben and up the coast to Alderton below Shingle Street.[96] South of Buckanay Farm the sea came over the shingle for 1070 yards in one place and for 733 yards in another. At Shingle Street there were two breaks in front of Oxley House of

92 Map A2727/2/4, Kirton–Woodbridge.

93 Map A2727/2/6, Ramsholt–Butley; Thomas E. Miller (1902–56) of Pettistree Hall Farm was NFU representative for Shottisham parish.

94 Norman E.E. Simper (1915–2006) of Manor Farm, Bawdsey, was NFU representative for the parishes of Bawdsey, Hollesley, Capel St. Andrew and Alderton.

95 A summary of all livestock losses is in the Appendix, below, p. 115.

96 Map A2727/2/6, Ramsholt–Butley.

44 and 108 yards and two breaks in the wall on the south side of Barthorp's Creek of 100 and 42 yards. The houses and the well known cafe of the house of Prichard Carr were isolated and marooned on the shingle bank; behind them was a flooded level of 500 acres and a raging sea in their front.[97] From Bawdsey to Shingle Street there were 112 acres of arable and 623 of grassland flooded.

Before turning up the River Ore, let us follow to the great shingle bank of Orford Ness with its Kings and Lantern marshes.[98] This never-never land of shingle, of terns and black headed gulls, is a prohibited area of the Ministry of Supply. When I flew over, its marshes were full of water from wall to wall. The marshes which have been grazed in recent years, have never been fully reclaimed [*Page* 103] from the condition in which they were left after a salt water flooding throughout the war. There were some breaches of which I only saw one opposite Orford.

Some months after the flood, I was talking to Mr. Walton of Woodbridge who hires the grazing of these marshes, of his livestock losses.[99] He told me one story of tragedy with humour. On the island is a latreen with multiple accommodation. In the surge of the flood, a heifer was borne over the marsh into this latreen, and was found drowned and sitting on a lavatory seat with its fore legs resting on the top of the door which was closed!

From Havergate, an island lying in mid-stream of the Ore, where there were 26 breaks, I learnt from a fisherman of Orford that he had seen between three and four thousand dead rabbits lying on the marshes: and he had also seen one alive on the wall! This island must have been completely awash and this one rabbit deserves commendation for saving itself.

Near the mouth of the Ore, there were 2 breaks of 55 and 218 yards which flooded the 480 acres of the Hollesley Bay Colony marshes.[100] Further up at Boyton there were 12 breaks and one at Butley Ferry. On the opposite side of the Butley Creek on the Gedgrave level, the wall took a nasty knock.[101] There were 7 large breaks, the smallest being 33 yards and the largest, 133 yards. The total length of the breaks was 595 yards with two top scours of 45 yards. The damage on this stretch of wall was comparable to the state of affairs around Green Point on the Deben: 200 yards was wrecked to the level of the saltings. Further up the level, there were 3 breaks of about 70 yards. In this area, 580 acres of good land was flooded for 3 weeks: about 200 of which was in arable production and had produced 9 quarters per acre of wheat and 20 tons per acre of sugar beet.

[*Page* 104] From Orford across the town marshes to Sudbourne right up to the foot of the hill below Iken church, the marshes lay like a vast sea, over an area into which so much thought and hard work had been put in the past six years.[102] Between

---

97 (Mary) Noreen Prichard Carr (1912–97) operated a cafe from a coastguards' cottage at Shingle Street from 1945. Her daughter recalls: 'because of lack of room inside, she made cakes and sold them and pots of tea from the sitting room window where customers queued up for beach trays to enjoy on the shingle beach beyond. Her strap line, yes she had one, was "Home made cakes of superlative quality" which was printed on a paper strip that lined the cake tins along with her name.' By 1953 the cafe had moved to Alde House, just up the beach, which had room inside for tables and chairs and a commercial kitchen (https://retrocooking.co.uk/about/).

98 Maps A2727/2/8, Gedgrave–Orford, and A2727/2/9, Sudbourne–Aldringham.

99 For a discussion of the historic practice of inland farmers renting marshes for summer grazing see Trist, *A Survey of the Agriculture of Suffolk* (London, 1971), pp. 57–60.

100 Map A2727/2/6, Ramsholt–Butley.

101 Map A2727/2/8, Gedgrave–Orford.

102 Map A2727/2/9, Sudbourne–Aldringham.

Haw Hill Sluice and Ox Sluice there were 2 breaks of 37 yards and the remainder of the damage was at Ferry Farm on either side of the bend of the river where the Ore changes its name to the Alde. There were 8 breaks totalling 72 yards and a further 10 badly damaged areas to the top of the wall totalling 38 yards. On the arable land in the extreme north east corner of the Ferry Farm opposite Slaughden, there was another muddle corner of boats, logs and other debris which came over the wall from the Aldeburgh side of the river. Below Yarn Hill there was a break of 5 yards. Further upstream below Iken cliff, the Dunningworth Hall marshes were again baptised with sea water with two breaks of 24 yards and further damage to the inside of the wall.

[*Page* 104A] To the west of Ferry Farm, Sudbourne over the ridge of the sandy hill which runs north to the Alde, lies Cowton farm which is approached by an overgrown track off Lamberts lane. Across the marshes to the west, stands the little farmhouse of Stannay: its buildings are sound, but the house has not been inhabited for some years and its roof is open to the sky.[103] Stannay lies in the centre of a block of marshes and like all dwellings found on the marsh level, it stands on rising ground – but it was not high enough to escape the flood. Behind the marsh, on the top of the sandy upland, lies Poplar farm Iken the home of Mr. Henry Fulcher who also farms Cowton and Stannay.[104]

At various points along the Suffolk coast, the annual rainfall is considerably less than the average. Storm clouds frequently threaten these areas, but pass over and the drought continues. To the west of the north end of the sandy upland dividing Cowton from the Ferry farm Sudbourne is one of these areas and the farmer's plight of one of dry marshes – an unusual state of affairs and an advantage in many ways. But like many others, Mr. Fulcher's land had an excessive unwelcome inundation of water at the time of the 1953 flood and I relate two instances of 'coping with much water' which will remain fresh in his memory for a long while.

In the cowshed at Cowton under the lee of the hill, there was eighteen inches of sea water and thirty-three cows to be milked. Here was a pretty state of affairs for a Sunday morning! It's a thousand pities that there is no recorded strip of the comments of Mick Mortell the Irish cowman as he bent down to each cow in turn and held the milking machine bucket! The water was too deep to put the bucket down on the concrete, and in any case a few inches of water would have floated the bucket on to its side. Then came the problem of cooling the milk, for the cowmen could not leave an [*Page* 104B] empty churn to fill under the cooler or it would have toppled over into the water. As soon as the machine buckets were full of milk, their contents were poured into a churn by one man, whilst another held the churn down in the water until the weight of the milk forced the churn to rest on the cowshed floor: the surrounding salt water completed the cooling in the good old fashioned 'pond immersion' method!

This was only one of two unusual jobs for a cold Sunday morning. Out at Stannay farm in the middle of the marsh, thirty odd yearling and two year old cattle were caught by the onrush of the flood at midnight and washed out of their yard, through the barn and across the marsh about two hundred yards until they were able to touch down on a sandy knoll, which stands a few feet above the main level. Looking across the flood in the early morning light, Mr. Fulcher saw his cattle marooned.

103 This is marked as Stanny Farm on the map, close to reference mark 4.

104 Henry Fulcher (1905–82) of Poplar Farm, Iken.

There was no boat, but somehow men had to reach the cattle and persuade them to swim to safety where they could be fed. It was a Suffolk horse which became a ferry. Four men of the farm, Keeble, German, Scarse and Walters stood by with Henry Fulcher, not yet knowing what was in store! One of the men mounted, whilst the rest hustled the horse into the flood for a swim over the fences across the marsh. But no! The horse panicked. It charged around in circles: it tried to jump and did everything except swim and one man couldn't keep it in the water. With one man up, another attempt was made, but it failed and then two mounted. This time they made a little headway but they couldn't keep the horse down in the water. Then all four of them mounted, with the last man sitting on the rump with his back to the third man, who held him, whilst he held on to the horse's tail. This did the trick, the horse had to swim and out they went over the marsh and landed up on the high ground. The cattle were driven into [*Page* 104C] the flood to swim and the 'old grey mare' surged ahead 'wi' Bill Brewer' and his three companions until they reached the dry ground below Poplar farm with all the cattle safe.[105]

On the opposite left bank of the Alde there was one break near Snape of 43 yards.[106] The famous humpback stone bridge stood up to a terrific ordeal as the sea poured under its arch to the ceiling and surged on to flood the level almost up to Blaxhall. The marshes south of Hazelwood Common took one knock in the wall for 70 yards. At Aldeburgh, the marshes had 13 breaks, whilst Slaughden was cut off from the town with a break in the shingle of 633 yards.[107] Further down the shingle, there were two gashes of 273 and 200 yards in Sudbourne beach. From the mouth of the Ore to the left bank of the Alde as far as Aldeburgh, there were 2192 acres of arable and 3553 of grassland flooded.

Beyond Aldeburgh as far as Benacre, there is only one stretch of sea wall defence and that is at Minsmere. For the most part the coastline is a shingle bank, sand dune or low cliff. At Thorpeness there were two breaks of 43 yards in the shingle. At Minsmere there was one break of 25 yards just [*Page* 105] south of the Coney hill cross wall which was itself badly broken for 96 yards.[108] Of the section of wall on which the tamarisk trees were growing at the time, some 300 yards had in recent years been almost buried in shingle, but the trees still flourished. For some weeks before the flood, the River Board had been rebuilding this length of wall, until the tide roared down under Minsmere Cliff and struck at it with all its might. The foreshore here is wide and deep in shingle and had been bulldozed into a protective bank a few years ago – but the tide came on. It threw down the shingle and cast away three parts of the height of the newly repaired wall and poured thousands of tons of shingle across the marsh for over a hundred yards: and still the tamarisk held – though alas! the trees have now been bulldozed away. This is a problem corner on Suffolk's coast for it has weakened to the tide on more than one occasion. The engineers scratch their heads and produce a new plan; will it include the old system of 'bent hills' which our grandfathers found helpful? [109] Yes! Since writing

105 A reference to the nineteenth-century folksong 'Widecombe Fair' about a man called Tom Pearce, whose horse dies after someone borrows it to travel to the fair with many friends. The chorus ends with a long list of the people conveyed by the horse: 'Bill Brewer, Jan Stewer, Peter Gurney, Peter Davy, Dan'l Whiddon, Harry Hawke, Old Uncle Tom Cobley and all'. It is the source of the humorous colloquialism 'Uncle Tom Cobley and all', meaning anyone and everyone.

106 Map A2727/2/7, Butley–Snape.

107 Map A2727/2/9, Sudbourne–Aldringham.

108 Map A2727/2/10, Aldringham–Dunwich.

109 Bent hills are dunes planted with marram grass.

this sentence, the sand and shingle has been planted with marram grass, both here and elsewhere along the coast.[110]

Between Minsmere cliffs and the River Blyth, there was one break through the shingle below Little Dingle Hill, in a position very close to the break which occurred in 1949 but south of Walberswick the sea swept over 833 yards of the shingle bank.[111]

How the Southwold harbour piers have stood as long as they have is one miracle, but how they survived this storm is another. As I looked over the Blyth in flood, I thought of Rennie, Cubitt and Walker and wondered if the river now had sufficient backwater to create a good flow?[112] I also thought of the old commissioners of the navigation: what had they [*Page* 106] been thinking?

On the river there were 26 major breaks and a number of places where damage occurred. The breaks totalled a distance of some 300 yards. The Southwold Corporation marshes of 130 acres had 6 breaks of 108 yards, the largest being 53 yards wide: whilst sand and shingle poured over the front as the sea swept over 470 yards of beach. The Charity, Tinkers and the Reydon marshes of 440 acres were flooded: the latter suffering 10 breaks in the wall. Further upstream, the Bulcamp, Angel and the marshes by Blythborough village were already subject to flood over 486 acres. The tide pushed on under the road bridge and ran up the valley below the Union and halted near Wenhaston. Down stream, it surged up Wolsey Creek and over the Halesworth – Southwold road and found a level all over the Henham marshes.

The sea did not make a damaging entry by way of the harbour, but came in over the beach on the north of the town below Easton Cliffs and followed a line north of Buss Creek over to the Reydon marshes: whilst some entry was also made over the beach front running south to the harbour. The worst of the damage to the river walls was done on the southern leg of Buss Creek where it faces the Reydon Marshes: here the tide was driven against the wall from the north west.

At Easton Broad there was a 30 yard break in the shingle bank, which is a poor defence at the best of times. The sea swept up the valley over Potters Bridge and on to Frostenden bottom.[113] At Benacre broad and at Covehithe, the tide pushed in a little further than usual: But at the Benacre sluice on the Kessingland level, it pushed in rather hard. The tide mounted the foreshore shingle and burst through the marram dunes by the coastguard station, making a gap of 85 yards over the top of the old sluice. For more than 100 yards on either side of the new pumping station, the ground level was completely altered. Hundreds of thousands of tons of shingle came over the dunes from the beach which bears the name of The Denes.

[*Page* 107] In the straight course of the Hundred river as it finishes at the old sluice, there was a band of shingle just under 300 yards in length: this was the course of the river, which was filled with shingle to its depth of about 10 feet. On the north side of the river, shingle and sand was washed 200 yards back into the marshes. The causeway between the water filled gravel pits, which leads down from Beach Farm to the sluice was almost eaten away in the centre and traffic had to be diverted

---

110 Marram grasses are dense, grey-green tufts (also known as bent grass), widespread around UK coasts. Their matted roots help to stabilise sand dunes.

111 Map A2727/2/11, Dunwich–Easton Bavants.

112 This is a reference to the civil engineers engaged in surveying Southwold Harbour and the general state of the River Blyth in the early 1800s. Trist writes about them and the committees that commissioned the work in ‘Records in the history of the Suffolk coastline’, see below, p. 39.

113 Map A2727/2/12, South Cove–Gisleham.

to the sluice by a track to the west of the pits. It was some days before this route could be used and all traffic of excavators and bulldozers to the sluice had to come down the coast from Kessingland. In the area of the sluice, I estimated that at least 5 acres had completely lost its former character by the washing away of dunes and the inroads of shingle. The complete alteration was finally made by bulldozers as they restored the level of the ground to enable traffic to move.

The pumping station was left in a devastated state with much damage done to the engines, but the concrete flume which conveys the water over the shingle, was not damaged. The connecting pipe between the pump house and the flume was torn out and crushed like a trodden cocoa tin. The crest of the shingle on the Denes right back to Benacre broad was flattened out and was subsequently restored to a height of 10 feet by bulldozers.

Whilst the pumps were out of action and the old sluice was still buried in shingle, the water in the Kessingland level piled up with the addition of fresh water, until it flooded back over the main Lowestoft road at Latimer dam, the lowest point in the level. The sluice was eventually opened and the river cleared of shingle at 2 p.m. on February 15th, during [*Page* 108] which time, there was frequent diversion of traffic to avoid the flooded road.

Again poor Kessingland of unhappy memory! During the war years, the level lay under fresh water as a defence measure.[114] In 1948 the Agricultural Executive Committee put in a fleet of excavators to restore the ditches. In the past few years, a temporary pumping station has been built and in the autumn of 1952 the water level in the ditches was sufficiently low for consideration to be given to plans for the reclamation of these 800 acres of marsh in 1953. Then came the night of January 31st and all the hopes of many years were dashed under a tidal wave. But Kessingland's level will again be green as her men will await in patience for the time to restore her marshes.

Against the tide we arrive at Lowestoft where to the north of the town the sea poured over the beach for 2,500 yards.[115] At Oulton Broad, water poured over the road from the inner harbour and one break on Oulton dyke flooded the Oulton marshes and part of the Barnby level. At Gorleston, the tide came over the beach by the Golf Club and for 1550 yards in front of the town.[116] There were 3 breaks on the Breydon Water.[117] The whole area of marshes from Yarmouth west to Burgh Castle were under many feet of water: the Herringfleet, Fritton, Belton, Burgh Castle and Humberstone Marshes comprising about 1500 acres were full of water. The actual damage in breaks and to scouring of the wall in this level was not so severe as elsewhere and the bulk of the water came over the top of the wall: but whilst it took a matter of a few hours to fill the level, it was six weeks before it could be pumped back into the Breydon water.

So much for the record of flood and damage to the river and sea defences. I think I am right in saying that on February 14th, a fortnight after the storm, all the breaks had been sealed to prevent further entry of the tide, [*Page* 109] with the exception of one at Levington Creek, one break with a deep back scour on the Aldeburgh wall and one in the Hazelwood marshes to the west of Aldeburgh.

---

114 A detailed discussion about the defensive use of the Suffolk Sandlings during the Second World War is provided in Liddiard and Simms, 2018.

115 Map A2727/2/14 Lowestoft–Corton.

116 Map A2727/2/16 Hopton–Great Yarmouth.

117 Map A2727/2/15 Fritton–Burgh Castle.

There were four stages in this operation of repair:

a) Emergency bagging protection to give sufficient height to withstand the tides of February 15th–17th: sand bags filled with the nearest mud or soil were set into the middle of the break and raised to within 2 foot of the top of the wall and in some cases to the full height.
b) Strengthening of work (a) by excavators filling up at the rear of the wall breaks. This work could not proceed until sluices and pumps had sufficiently lowered the water level to enable the excavators to crawl along to positions to work.
c) The permanent repair of the breaches
d) The permanent rebuilding of the remainder of the damaged wall.

The restoration and rebuilding work has been proceeding throughout the summer of 1953. The marshes have surely never 'looked' so safe as they appear to-day, with their new high walls raised like railway embankments along the rivers. Before the flood the average height of these river walls was 12ft. O.D. with a 3 foot top and a base width varying from 20–30 feet.[118] The width of the 'folding', that is the land between the foot of the back of the wall and the delph ditch, varied from 16–25ft. The width of the delph ditch varied considerably from 8–12 feet with a variable depth up to about 5 feet according to the standard of its maintenance. In many places the height has been raised by 2–2½ feet. The new width of the top of the wall is 6 feet. The 'folding' [*Page* 110] varies from 30 to 45 feet to the edge of the delph ditch, which itself has a width range equal to the 'folding'.

The emergency work has been torn out by the excavators and bulldozers. The top soil over the marked out new delph has been put to one side, whilst the excavators have dug out the clay from the new delph ditch which is piled up on one side to be later bulldozed up to the foot of the new wall and positioned by excavators. As the raw wet clay settled on the new wall, the top soil was moved and spread over the top and the sides of the wall. This more friable soil runs into the crevices which are formed as the clay dries and shrinks and assists consolidation: and will later form a surface for reseeding or turfing to grass as a protection against the weather and the tide.

To tidy up the record, I should include some statistics which need no elaboration. One of the most important records taken by the County Staff of the National Agricultural Advisory Service, is the information relating to the duration of flooding. These records were important in dealing with advice in connection with soil sampling for salt content.

118 Ordnance Datum, the benchmark from which tide height is determined. OD marks height zero on maps in Britain, so '12ft OD' means 12 feet (3.5 metres) above mean sea level.

East Suffolk: Duration of Flooding

Time		Acres Arable	Acres Grass
0–12	hours	123	7
12–24	"	274	258
1–2	days	170	368
2–4	"	261	881
4–7	"	327	1340
7–10	"	522	1435
10–14	"	1013	2150
14–21	"	1267	3880
Over 21	"	77	6156
		4034	16475

Livestock Losses

Cattle	146
Sheep	303
Pigs	86
Poultry	1289
Horses	11

[*Page* 111] There were 44 cases of damage to farm buildings, 5 of which were serious. The majority of the farm buildings on the Suffolk coast farm are off the low level of marsh and consequently, other losses in store, were fortunately comparatively slight. There were 42 tons of fertilizer flooded, 35 tons of bulky fodder, 56 tons of feeding concentrates and 50 tons of mangolds and potatoes in clamp. In addition, a number of farm implements and equipment was flooded, the majority of which was rescued in time to restore it to working order, although a number of tractors needed a substantial sum of money spent upon them to put them back in working order.

On the 14th February, the D–day for sealing off the water before the onset of the new high tides, the total force of men and equipment engaged, was 1946 civilian workers (excluding engineers, technicians or other staff from various organisations), 1765 army and R.A.F. men, 18 bulldozers, 14 pumps, 17 excavators, and 32 river craft.[119]

[119] There was a general appreciation for the scale and success of this operation. An editorial in the local newspaper assured readers that even the Home Secretary, Sir David Maxwell Fyfe, had been wonderfully impressed by local people rallying to aid those afflicted and ‘the thousands of volunteers and Servicemen who have slaved on the river walls to fill the gaps’. It concluded that, ‘It has been a race against time under abominable working conditions in which snow, hail and rain have done their best to impede operations. It looks as if these super human efforts have gained their just reward and that the tides have been defeated’ (*EADT*, 17 February 1953).

The story of this great flood would not be complete without some mention of what happened in the coastal towns. For this information, I am drawing on my own impressions from visits to the towns after the disaster and from reports in the East Anglian Daily Times which kept the public very well informed from day to day.

At Ipswich, soon after 11 p.m. on January 31st, the water from the dock broke across Stoke Bridge and into Bridge Street, Commercial Road, Quay Street, Fore Street and into the area of Princes and Bath Street where the water was 4 foot deep by 1 a.m. Doors were burst open, walls collapsed and cellars were flooded. Considerable stocks of animal feeding stuffs and fertilizers were flooded in warehouses. There was no loss of life but many people had to be evacuated from their homes. In spite of a distance of about 12 miles from [*Page* 112] the open sea, Ipswich took its share of the flood which surged up the wide estuary of the Orwell. At Woolverstone nearly every yacht suffered considerable damage: whilst at Pin Mill, houseboats were sunk and some of the houses flooded to a depth of 5 feet.

At Felixstowe there was tragic loss of 41 lives and considerable damage to property. The sea came over the Dock area and from the Orwell at Fagbury Pt. where the wall was severely breached at one point and the entire length of the top of the wall along the old oyster beds, was washed away. The sea swept up Langer Road and flooded the entire neighbourhood to a depth of 5 to 6 feet. Some warnings were given, but the onrush of the water was so rapid that many people were caught in their beds.

The scene in the vicinity of the Beach Station gave some idea of the fury of water and storm. Dozens of caravans were floated across several hundred yards to be lodged like matchboxes on the side of the railway embankment. On the opposite side of the road, Mr. Simper's Station Farm was badly hit. His entire dairy herd of 26 were drowned, his buildings badly damaged and all his crops and grass were under several feet of water. A builder's yard alongside was in utter chaos.

In the Langer Road, the pre-fab houses were moved like packs of cards. The tragedy and the chaos will long be remembered, and Felixstowe can remain proud of her people who rallied to the assistance of those in distress.

Whilst the surge smashed the walls of the Deben, it came over and through the marsh wall at Woodbridge. It overtopped the quay by two feet and flooded quayside cottages and left a tide mark of 2 feet in the Boat Inn and filled [*Page* 113] its cellar to the ceiling. The screen end of the Cinema was flooded to a depth of four feet. There was loss of stocks in the Tide Mill and considerable damage done to boats and electrical equipment stored in Everson's and Knight's boat yards. The Eastern Counties Farmers Co-op. warehouse was flooded. Beyond the town, the Melton Corn and Coal Company lost about a 100 tons of coal and coke which was washed out of their yard.[120]

Below the Brudenell Hotel at Aldeburgh, the determined tide had treated the great shingle barrier with contempt and pushed its way through into the marshes, strewing thousands of tons of shingle to sprawl over many acres. For the first time within living memory, the sea found its way into the High Street where shopkeepers had to remove their stocks to safety and boats could be seen tied up to lamp posts. Cattle were lost on the marshes at Aldeburgh Hall, whilst many smallholders lost pigs and poultry. Slaughden quay was swept clean of its sheds and big and small

120 Reported in *EADT,* 2 February 1953.

boats were thrown in all directions. Yachts were torn from their moorings and suffered much damage.

When I first saw Aldeburgh after the disaster, on February 19th, the houses in the neighbourhood of the Brudenell Hotel still had sandbagged doorways for fear of further high tides. On the shingle bar there was an incredible scene of bulldozers and excavators trying to restore order. The marshes were still full of water for one breach with a deep under scour had not yet been sealed. Outside the river wall, all sizes of boats lay cast up high on the saltings.[121] Here there was no loss of life at the time of the flood, but an unfortunate tragedy occurred on the 17th when Daniel Mann, an Aldeburgh lifeboat man lost his life. Two fishing boats, one carrying sandbags and the other ferrying soldiers on the flood inside of the walls, were sucked through the breach by a strong ebb tide. The two [*Page* 114] boats collided and turned over and Mann was carried away by the tide.[122]

On February 27th the Aldeburgh Council met to review its position.

[1] [*footnote in text*] "Satisfaction was expressed at the way the town's sea defences stood up to the tremendous test imposed upon them during the recent gales and floods of January 31st.[123] The report of the Sea Defence Committee said the Council owed a great deal to the sound work and design put in to them.

"The Committee's report stressed that prior to the storm they had expressed concern over the long delay in carrying out work on the River Board's sea wall and that the Town Clerk had been instructed to inform the River Board that they would be held responsible should the Corporation marshes be flooded, with consequent loss of rental. The Committee's fear had been justified and the loss to the Corporation could not yet be assessed". The report went on to state that as a result of the new design some divergence of opinion had arisen with regard to the contract for this work.

"The view of the Committee was that the new design could be held to be within the variations clause of the existing contract but the River Board did not agree with this opinion". The question had therefore been referred to the Minister of Agriculture and Fisheries by the Clerk of the River Board.

"Dealing with the River Board section the report said the most obvious damage was to the groynes which were under-scoured to a depth of six to eight feet in places on the upper beach and badly broken up between the end of the concrete blocks and

121 The long-term policy about sea walls was frequently discussed in the House of Commons and the situation at Aldeburgh was used to advance the case 'for the establishment of some overriding authority, operating through one Ministry' to take responsibility for flood defences. Colonel Harrison, MP for Eye, claimed that after the 1949 storm the borough of Aldeburgh spent a lot of money to 'erect a first-class sea wall which was 99 per cent successful' but nothing was done elsewhere, so the town remained vulnerable. Though they had made their own sea wall safe, in 1953 sea water came in to flood the people in Aldeburgh because of weak defences to the north and south (*EADT*, 20 February 1953).

122 See *EADT*, 18 February 1953; William V. (Billy) Burrell was awarded the British Empire Medal for his rescue of the survivors in this incident. When the soldiers and fishermen were thrown into the water, Burrell realised the danger and quickly took charge. An extract from the citation explains that, 'One of the fishermen had already persished in the water and Burrell hurried on to the aid of the remaining four men who were pulled into his boat and eventually landed and taken to hospital. The whole incident was quickly over but there is no doubt that the four men who were rescued owe their lives to Burrell's speed of thought, action and courage' (*Supplement to The London Gazette*, 28 April 1953, p. 2332).

123 [*Trist's footnote reads*] East Anglian Daily Times report of March 2nd., 1953.

the end of the river wall where a bad break through in the beach let in a big volume of water to the marshes to reinforce the great quantities that surged over the wall".

The shingle bank of three and half miles between Dunwich and Walberswick [*Page* 115] is a vulnerable part of the coast. As in the 1949 tide, the sea again came over the shingle at the Walberswick end for a considerable distance.[124] To the north of the old Corporation marsh wall up to where the huts stand on the higher ground of the sand dunes, the beach is low and an inviting entry for the tide. It poured back into the Dunwich Creek and over into the gardens. On the beach below the little arable field on the high ground, the waves had exposed the sandy soil of the mainland. The shingle bank was clearly flatter and more shingle had been driven back into the marshes towards the Creek.

Mr. H.A.P. Jensen* [*footnote in text*] of the Nature Conservancy Board writing in 'Weather', a publication of the Royal Metrological Society, recording erosion after the floods, stated that "The natural dune, shingle and cliff barriers have withstood the brunt of the storm remarkably well, but at least a 10 yard depth of outer dunes have been cut clean away over long stretches of coast.[125]

"The levels of many beaches have been lowered by a foot or more, and the crests of shingle ridges have moved inwards by up to 20 yards". "At Covehithe in Suffolk, losses may have been between 10 and 30 yards in places, while at Dunwich erosion has occurred on the cliffs which have been relatively untouched in recent years". "Coasts protected by engineering works have been relatively free from major breaches".

Southwold, suffered both tragic loss of life and considerable damage to property. There were few people who had a night's sleep, for many rallied to the rescue of those caught by the tide which swept in over the front and to the north of the town under Easton cliffs. The sea was coming into the town by 7 o'clock and before [*Page* 116] midnight Southwold was an island. A van attempting to drive through the flood at Reydon bridge, was washed off the road into the river. The only approach to the town on the following day was by ferry at Reydon bridge which was operated by Mr.Winter with the aid of a tractor.[126]

On the sea front along Ferry road, the scene must have been terrifying. The shingle bank of the shore lay about eight feet above the road which was flanked on one side with bungalows. As the tide burst over the bank, it hurled eight to ten feet of shingle into the road below. Four bungalows completely disappeared and sixteen others were severely damaged: some looked as if huge chunks had been eaten out of their corners. About thirty unoccupied summer bungalows were swept out to sea. One American family were rescued from the roof of their bungalow after it had been driven about 400 yards out into the marshes. The family Sorick were lucky people and in the circumstances one can only imagine that at times, some people have great presence of mind. They were rescued, after two trips in a boat, by Mr. ~~Mobbs~~ F.J. Mayhew who was a longshore fisherman and former lifeboat coxwain, and Mr. E.W. Stannard who accompanied him.[127]

124 Map A2727/2/11, Dunwich–Easton Bavants.

125 [*Trist's footnote reads*] *From the East Anglian Daily Times April, 1953.

126 John Winter later recalls his memories of the flood, including seeing an ambulance washed away and someone carried from their home on a stretcher made from a door (www.southwoldmuseum.org/thesea_1953Floods.htm).

127 Mayhew was awarded the British Empire Medal for his part in the rescue. The citation reads: 'During the first night of the floods, Mayhew was responsible for saving the lives of two women, a baby and

When I first saw Ferry Road after the floods, I thought it looked like the biggest chaotic muddle I had seen in any of the coastal towns. A road full of shingle seemed a small matter which was being corrected – but everywhere there was the litter of personal belongings and property. Chairs and tables, doors and window frames lay in heaps along a line as the tide had left them. Boats had been washed over the top of the beach and lay upside down on the marsh. It was an indescribable muddle which only the fury of raging water can create. I felt very sorry for Southwold; but under the able leadership of her Mayor, Major General R.J. Mackesy and the Town Clerk Mr. H.A. Liquorish, Southwold was quickly about her business of restoration.[128]

[*Page* 117] Some months later, I walked the Corporation Marshes with the Town Clerk. This was the one place where I had no sympathy with Southwold. The flood with its attendant poison of salt water was the best thing that ever happened to these marshes, which had clearly been in an appalling state before the sea finally ruined them.[129] Now they must have some attention in a big way. But Southwold is not alone in this respect, for these remarks could equally be applied to many areas of these coastal marshes.

The cliffs at Easton Bavents suffered further erosion and there was evidence of recent falls on to the beach. About ten days before the floods, the Lothingland R.D.C. had discussed the position of the eleven houses which were in danger of falling into the sea. Erosion is moving fast on this part of the coast. One occupant recently spoke of an advance of 10 yards in two years; whilst Mr. Boggis of Bavants Farm estimates that he has lost about 15 acres of his land along the cliff, in the past 50 years.[130]

Later, in February[1] [*footnote in text*] the Lothingland R.D.C. on the advice of its Sea Defence Committee decided that they could not undertake a defence scheme to prevent further cliff erosion.[131] It was stated at the meeting that the cliffs had eroded 20 feet on the night of the storm.

three men. This involved making two trips in a small rowing boat across flooded marshes under very hazardous conditions. Each trip was fraught with considerable danger as the boat had to be navigated in the dark across an area interspersed with submerged fences and walls which could easily have holed and sunk the boat. Gale conditions whipped up the water and the rescue called for a high degree of gallantry and skill.' Stannard was awarded the Queen's Commendation for Brave Conduct, 'for helping to rescue the occupants of a flooded bungalow at Suffolk' (*Supplemement to The London Gazette,* 28 April 1953, pp. 2333–34).

128 Major-General Pierce Joseph Mackesy (1883–1956), a veteran of both world wars and mayor of Southwold four years in succession from 1949. He was married to the romantic novelist Leonora Starr (Memoirs, *The Royal Engineers Journal* 70, No. 3, 1956, pp. 295–97); Harold A. Liquorish (1900–60).

129 Trist was an advocate for the 'proper management' of all land. He found neglect abhorrent, and a detailed discussion of the merits of land rehabilitation and reclamation are provided in his *Land Reclamation* (London, 1948).

130 Herbert C. Boggis (1872–1971), dairy farmer at The Warren, Easton Bavents, bought the 400-acre estate in 1925. According to Herbert's great granddaughter, his vision was to turn the land into a leisure village of 80 houses, along with bowling greens and tea gardens. His plans, drawn up by the famous architect Patrick Abercrombie, a founding member of the Council for the Preservation of Rural England, were thwarted by the outbreak of war (*Suffolk Magazine,* 11 July 2019). Ninety years later, Peter Boggis, born on the estate in 1932, estimated that half of the land had since been lost to the sea. Boggis, an engineer, was notorious for battling to build his own sea defences after authorities withdrew support for protecting the cliffs at Easton Bavents (*The Guardian*, 2 April 2015).

131 [*Trist's footnote reads*] 1. East Anglian Daily Times Report of 19th February, 1953.

'The wildest night of disaster the port had ever known' was the opening comment of the scene at Lowestoft as reported by the press[2] [*footnote in text*].[132] 'The whole of the North Denes guarded by the new £500,000 sea wall, had been inundated as the waves soared clean over it, flooding the gas works, where the furnaces had been put out by the water'. 'Overnight, the borough had been suddenly plunged into darkness by the failure of the electricity supply after the local works had been flooded'. 'In the harbour and at the swing bridge the tide reached almost double the usual rise. Vessels moored to the quays of the docks were lifted until they were riding, in some cases, against the roofs over the fish markets [*Page* 118] and, with the bollards completely out of sight under water, moorings slipped'.

'Three drifters were swept ashore on to the Northern Extension, while fish boxes and tree trunks were carried as far as the main street. Hundreds of barrels of pickled herrings were washed all over the Denes and streets in the Beach district'. 'The town itself was completely cut in half just before high water time on Saturday night, when the seas suddenly surged over the harbour walls across the main road, both north and south of the swingbridge'. 'At Oulton Broad, water from the Inner Harbour went straight through a big grocery store and restaurant. A hotel and other shops were inundated with three foot of water'.

The might and power of destruction of flood water could be seen on the north wall of the Hamilton Docks at Lowestoft. Here, a gigantic wall of concrete was forced out of the vertical by the pressure of the flood.

If South Town, Gt. Yarmouth had not been partly protected by the railway embankment, the flood might have caused greater disaster. From Burgh Castle to South Town, more water swept over the river wall than broke through. It lay for a period of six weeks whilst pumps worked night and day to return the flood to the Breydon Water.

Mr. Vincent of Humberstone Farm must have known as much about the Yarmouth floods as anyone.[133] His farmhouse is more than a mile into the marsh and stands on the same level. The farm was completely inundated, and with thirty odd cattle to save, he had one way out and that was along his narrow farm road which is flanked on either side by a deep ditch. Walking up to his chest in water, he drove his beasts along this flooded road and only lost one!

All this is but a brief picture of this great disaster as if affected East Suffolk. The loss of human lives has left its tragedy, and the destruction [*Page* 119] to property and land will cost millions calculated over the years for which it will have an effect.

The story of the rehabilitation of the land will be long and its restoration will take some years. There is no immediate antidote for the poison of salt left in the land. The first task was to release the salt water from the ditches and this has been assisted by excavating them to encourage a quick flow to the sluices in the river and sea walls.

A large percentage of the marsh grassland has been completely wrecked and contains a high percentage of salt which only time can heal. When the salt has lowered to a safe degree, the marshes must be reseeded to new grass.

As I finish this record in August 1953, a large part of the marshes up and down the coast are bare of grass and are a mass of salt tolerant weeds. The arable land lies idle, and out of 4,000 acres flooded, only 500 have borne a crop in 1953, with a prospect of less than half of the normal yield.

132 [*Trist's footnote reads*] 2. East Anglian Daily Times Report of 2nd February, 1953.

133 W.H. Vincent (1913–2001), Humberstone Farm, Cobholm, Norfolk.

The new sea defences go up at a great pace. The river has never before witnessed such giant barriers. Let us hope that the next fury of the tide will be frustrated in its attempt to surmount the walls, and that it will leave the marshmen to go about their new husbandry with confidence.

RECORDS IN THE HISTORY OF THE SUFFOLK COASTLINE

(largely built up from the archives in the Ipswich Borough Library, extracts of which have been published in the *East Anglian Daily Times*.)

[*Page* 120]

~~Chapter 5~~ [Chapter] 4

"<u>Records in the history of the coastline</u>"[1]

In this chapter, it is only intended to give the reader a brief history of the problem of the defences of the Suffolk Coast line. A comprehensive record would entail a vast amount of research among old books and documents for which I could find insufficient time: and I fear that having gone to such length, the detailed result might only be of interest to a limited few. Therefore my attempt will comprise a few snatches from records, sufficient to be of interest to the general reader; more especially to those who know the Suffolk coast.

With little imagination, it is obvious that the present coastline is 'comparatively' new. An ordnance survey map of the coast, if taken every ten years would continue to show that the sea is a master in its own domain and bent on further conquests. For hundreds of years the sea has been gnawing at the coast in some places, whilst in others such as Minsmere and Easton Bavants, the river mouths have silted up. A Minsmere man has taken up this advantage and enclosed the land from the sea – but its enclosure is for such periods as the sea decides in its periodic combined revolution with the north west wind.

The low lying marshlands are comparatively new additions to our land mass and have generally been enclosed from the rivers and the sea during the past four hundred years. In the 16th century, the Dutch were already busy on sea defences and many of their early engineers came over to this country to advise and assist with the work of building sea walls – and still do.

All the river walls which enclose the marshes of the Stour, Orwell, Deben, Ore, Alde, Blyth and the Breydon Water are therefore protecting land which has been won from tidal waters. Throughout the centuries, these marshes [*Page* 121] have been won and lost and then regained. Some of them to-day are temporarily lost to the tide, and in spite of the urgent need for all the acres of food producing land, the problem of financing these works must loom its critical head and wag words of economic wisdom.

But there are several ways of looking at the argument. The coastline of East Anglia is one enormous headache for engineers who have been endeavouring to solve the problem for many years. The question of the physical and financial difficulties in building a tidal river wall is one thing, whilst the problem of stabilizing the defence of a long shingle shore is yet another. It is not unusual for a high tide to move two feet off the crest of a shingle bank in one tide and then within a brief period, put it all back tidily in the same place.

The history of Suffolk's coastal defence is therefore largely concerned with one long battle with the wind and sea. Although our intensive farming period can only be said to cover the last 70 years, farming was rapidly gaining momentum in technique from the latter part of the 18th century. But before this period, the East Anglian coastal farmer must have recognised the difficulties of his rather dry climate, and it

1 This document was written by Trist from his research and the footnotes are largely his, indicated by [*footnote in text*] where the marker is inserted and [*Trist's footnote reads*] within the footnotes. His idiosyncratic numbering system has been preserved.

is certain that marsh and salting grazings were considered a valuable adjunct to the light sandy hinterland so subject to drought.

In addition, Suffolk had trade which carried men in ships, and ports and waterways were important for navigation. For this purpose it was necessary to have some control over the flow of inland water as well as defence against the sea. Down the centuries, farm produce and other merchandise left from a number of ports on the Suffolk coast – to-day this trade is disappeared except at such ports as Ipswich, Lowestoft and Yarmouth. Not only are wool and wheat now transported overland, but the ports of the past to [*Page* 122] which I refer are no more! There was a time when Suffolk built many ships for the Royal Navy: and less than fifty years ago, much barge traffic plied up and down the river. To-day they are almost forgotten and the surviving bargemasters are probably the last of their race.

The river walls were then serving purposes other than enclosing land. Small quays were built for the barge and ferry traffic and some of the many footpaths that cross the heath country between the rivers, lead down to water where a boat across the river provided quicker transport than a long detour around the head of the river, over bad roads. But I must say no more of the river and confine myself to the land over which men tread in their way to the sea.

There have been laws governing sea defence and the general protection of waterways, at least dating from the early part of the 14th century. In the reign of Edward III (1327–77) and again in the time of Henry VI (1422–61) statutes were passed governing sea defence; but it is unlikely that much serious sea defence was accomplished before the latter part of the 16th century. In 1532 the Statute of Sewers was passed. There were five main parts in the Act of which the first three are significant for our purpose. To study the Act, I turned to a paper[1] [*footnote in text*] of one Robert Callis which was published in 1622.[2] Legal language is often difficult and ponderous to follow, for precision and accuracy to cover every contingency is [*Page* 123] necessary in legislation: nevertheless, let us see what we can make of this statute which was a forerunner of modern acts to give protection to farming and facilities to shipping. The preamble reads …

> 'Knowye. that forasmuch as the walls ditches banks gutters sewers gotes[1] [*footnote in text*] calcies[2] [*footnote in text*] bridges streams and other defences by the coast of the sea and marish-ground lying and being within the limits … in the county of the same, by rage of the sea flowing and reflowing, and by mean of the trenches of fresh water descending and having course by divers way to the sea, be so disrupt lacerate and broken and also the common passages for ships ballangers[3] [*footnote in text*] and boats in the river streams and other floods within the said limits or in the borders or confines of the same by means of silting up erecting and making of streams mills bridges ponds fisgarths milldams locks hebbing[4] [*footnote in text*] weirs hecks[5] [*footnote in text*] and floodgates and the like letts impediments and annoyances be letted and interupted so that inestimable damage for default of reparation of the said walls ditches ….'[3]

2 [*Trist's footnote reads*] 1. The reading of the famous and learned Robt. Callis Esq. upon the Statute of Sewers, 23 Hen. VIII C.5 as it was delivered by him at Grays Inn in August 1622' – by William John Broderip.

3 [*Trist's footnote reads*] 1. Gotes: 'engines erected and built with perculleses and doors of timber…. for draining the waters out of the land into the sea'. i.e. sluices; [*Trist's footnote reads*] 2. Calcies: 'a calsey or calsway is a passage made by art of earth gravel and stones on or over some high ground or common way…' i.e. a causeway or hard; [*Trist's footnote reads*] 3. Ballenger or balinger: a small sea-going vessel probably without a foc'stle; [*Trist's footnote reads*] 4. Hebbing weir: a wier for catching fish on the ebb tide; [*Trist's footnote reads*] 5. Heck: a grated framework for trapping fish.

So far, we learn that walls, banks and ditches were in being, but 'be so disrupt': and that waterways were silted up and all was in 'default for reparation'. The preamble continues with urgency '... and yet is to be feared that far greater hurt, loss and damage is like to ensure, unless that speedy remedy be provided in that behalf.' 'We therefore for that by reason of our dignity and prerogative royal we be bound to provide for the safety and [*Page* 124] preservation of our realm of England have assigned you and six of you to be our Justices to survey the said walls streams ditches ....'

It is quite clear that the importance of the necessary work was recognised and it is interesting to follow through subsequent records of meetings of the Commissioners in the 18th and 19th centuries and see the way in which these bodies considered the importance of their commission. Callis in speaking at Grays Inn quotes from a Mr. Camden to stress his own views of sea defence ... '"quod insula Britannia avida in mare omni ex parte se projecit ..." that our realm with sea on all sides could not be safe without these provident laws'.

The Commission appointed 'true keepers, bailiffs, surveyors, collectors and expenditors ... and to hear the account of the collectors for the receipt and laying out of the money that shall be levied'. They had power 'to tax assess charge distrain and punish ... after the quantity of their lands ...' and debtors could be detained and punished 'by fines pains or other like means after your good discretion'.

This statute meant business! For not only were the commissioners given a heavy responsibility, but they were armed with a 16th century form of fierce Defence Regulations, in as much that a defaulter could be fined or perhaps lose his ears at the pillory!

Callis pointed out that 'the ownership and property of the sea bank and banks of great rivers be to them whose grounds are next thereto adjoining'; 'and the trees, grass and other things thereon growing belong to the owner of the soil: but the use of the banks is common to all the King's liege people ...' 'and the owner of the soil cannot justify the digging of casting of them down' ... to hinder the people's right. Private river banks erected to protect men's private grounds were outside the Act and Callis comments that if neglect causes nuisance, 'then there shall be action'.

To-day, most river walls are the responsibility of the newly formed river boards, formerly the catchment boards, but neither banks of the River Orwell in [*Page* 125] Suffolk have ever come under the jurisdiction of a board, solely because the owners preferred to maintain their independence of responsibility; once a pride and now a source of heavy financial burden. And if there are breaches made by the sea or a continuation of lack of maintenance, no action is taken, as suggested by Callis.

The Commission in setting up local commissioners of sewers, charged them to 'see' for themselves and they were required to view the defences 'noteing faults and defects and by conference with carpenters and masons ...' to watch over sea encroachment. 'If any wall, bank, river ... or other defence be defective by neglect of sufferance of such as should repair the same, the commissioner of sewers are to enquire by jury in whose default the same happened'.

Where defective walls required timber for the work, the Commissioner's men were authorised to give notice to the owner of trees near the work, that they would enter and take the timber required, for which they would pay a reasonable price. They could take six trees and were 'to enter at seasonable times without doing damage to the said land ...' As an ordained right, it would appear reasonable in those days, for transport was yet in its infancy and all roads were in a shocking state compared with to-day's standards. The problem of carting timber over marshes to river walls is often bad enough without multiplying the burden and the cost of hauling timber from long distances: but at that time, any other arrangement than that ordained, would have put a heavier tax burden on the landowners.

Their terms of reference were clearly defined in the Statute and we shall later hear from other documents, of the serious way in which the Commissioners reported to their committee and recorded their minutes. This statute is probably the foundation of our legislation in regard to sea a tidal river defences.

[*Page* 126] The realisation of the necessity for this legislation is significant in the history of our agriculture, for whilst its provisions were concerned with coastal protection, interior drainage channels of all kinds were taken into account: and this work is of fundamental importance to farming whatever the type of land. This Statute is also significant for a period in our history, for its advent is at a time when the Renaissance was being ushered in. It was the beginning of 'modern times' when we start to see the last of the old feudal lords and a new set of landowners drawn from the wealthy middle class. These new lords were men interested in peace, so that trade could prosper without interruption of war and it was in their interest that the disturbance created by rival factions of the Roses, should cease.

The Renaissance did not flower in isolation, its seeds spread and improved strains sprung up. It was a period of beauty, of colour and expression, but like most ages it did not overcome the vulgarity and cruelty of its own making. England prospered in discovery, in mastery of the sea and created heroes, poets and playwrights. Although farming was still in its infancy, the world of literature gave rise to thought and improved methods already exercised the minds of a few.

In 1523, Fitzherbert of Norbury is credited with the first book on farming, published in English. 'The Boke of Husbandry' is an apt illustration of Renaissance, for Fitzherbert points out that farming land is not a thing just to be left to workers. He was a pioneer advocating farm management and called attention to the necessity for the interest of the master, whom, he said, had to learn by practical experience – 'it is the best way that even I could prove by experyence, the which … have assied many and dyvers wayes and done by dyligence to prove experyence which shuld be the best waye'.

In the same period when the Statue of Sewers was passed, we have Thomas Tusser whose dates are queried, but believed to be between 1524 and 1580. [*Page* 127] Tusser's first edition was called 'A Hundreth Good Pointes of Husbandrie', all of which were in verse. The book subsequently grew and the third edition contained five hundred points of good husbandry, together with notes for the housewife. Here we see thought, the verse and the colour of the period coming out in a new education for the land.

> "Who breaketh up, timely, his fallow or ley,
> sets forward his husbandry, many a way;
> This tilth in atilture, well forward doth bring
> not only thy tillage, but all other thing.
> "Though ley land ye break up, when Christmas is gone,
> for sowing of barley, or oats thereupon;
> Yet haste not to fallow, till March be begun,
> lest afterward wishing, it had been undone."

Tusser says much in a few lines and in his advice on the sorrows that befall impatience on the land, it is clear that he knew of the evils of water and well he might, for he lived hard by the banks of the River Stour. Tusser farmed at Cattawade, now occupied by Mr. Frank Keeble of Brantham Lodge. 'Alas! my friend, whatever the estuary was like near the walls in your time, it must now be very different. Mud islets have now appeared by the aid of rice grass (spartina) which was planted in 1923 by Mr John Keeble'. 'But some of your land is still farmed, Thomas, and the name of your hamlet is perpetuated in honour in Suffolk. You will know of the Suffolk Punch and for some years, many of the champions of this breed have borne the prefix of Cattawade'.

No doubt many of the events leading up to this early coastal protection act were connected with the destruction by the sea, of 'defences' on the coast and in [*Page* 128] the history of Suffolk, none in more memorial than that of Dunwich.

By the later years of the reign of Henry VIII Dunwich was already reduced to one quarter of its original size. In 1570 there was a terrible storm[1] [*footnote in text*] when the sea claimed several churches and again spelt ruin to the harbour.[4] Queen Elizabeth I in 1578, considered the losses of the people of Dunwich and made them a loan which was raised by the sale of 'belles, lead, Yron, glasse and stone wherof are vallued at threscore syxtene poundes, eyghtene shillings and fower pence', from Ingate church in Suffolk'. The plight at the time is described in the original document … 'Whereas we are crediblie enformed, that the Queene's Majesties Town of Dunwytche in the Countie of Suffolk, is by rage and surgies of the sea, daylie wasted and devoured, and the haven of her highnes said Towne, by diverse rages of wyndes continually landed and barred, so as no ships or boates can either enter in or oughte …'

Year by year, the wind and the sea gnawed as the sandy cliffs and the foreshore, until the trade of the town and prosperity was undermined to such an extent that a once prosperous city became a burden for charity from the Crown.

Another great storm in 1608 destroyed the road to the beach which at that time may have been two miles or more out to sea from where the present village lies. In 1677 the sea invaded the market place and appears to have firmly won ground from Dunwich, for in 1680 many buildings were taken down by the inhabitants before the sea attacked. In 1702 a storm made further inroads on the Town and the church of St Peter was taken down: and by 1715, the gaol was undermined.

[*Page* 129] Gardner records a terrible storm in December 1740 when a gale blew for several days causing high seas which broke down banks and flooded many marshes. "The sea raged with such fury, that the Cock and Hen hills, which the proceding summer were upwards of forty feet high, and in the winter partly washed away, this year had their heads levelled and their bases, and the ground about them so rent and torn, that the foundation of St Francis' chapel, which was laid between the said hills, was discovered." Foundations of old houses and churches were laid bare and several skeletons were "on the ouze divested of their coverings". Gardner records more furious tides in the winters of 1746 and 1749. When the Rev. A. Suckling wrote in 1848, he recorded that "no considerable encroachments of the sea have taken place in Dunwich for about seventy years past" so that three generations of its inhabitants appear to have avoided a sudden bolt from their houses during that period. This seems to be borne out by the population which was recorded as about 100 in 1754, which by 1841 had risen to 237.

4 [*Trist's footnote reads*] 1. Rev. A. Suckling: History and Antiquities of County of Suffolk, recorded from Gardner's History of Dunwich. 1754.

Thus the seat of St Felix, who spread the gospel of Christianity throughout East Anglia, has been wasted over the years and the ravages of the sea continues. The few houses of Dunwich to-day huddle together on higher ground above the beach, whilst the sea o'ertops the great shingle bank which runs up to Walberswick and floods the remaining marshes, the Dingle and the Reedland, on those occasions when a furious gale whips it up. In 1877 and 1897 there were high tides which flooded many marshes in Suffolk and no doubt the Dunwich marshes suffered at the same time.

Arising out of the Statute of Henry VIII, we find excellent records of the outcome of this legislation, in the minute books of the precursors [*Page* 130] of the present day River Boards. On the 16th of April 1779 a Court of Commissioners[1] [*footnote in text*] held at the sign of the Fox in Hollesley, considered their duties.[5] They first appointed Richard Wood of Melton as Clerk bailiff and collector to them and agreed he should be paid his expenses and a salary of 6d. in the £ of moneys collected. Two surveyors were appointed at a salary of 1 shilling in the £ collected.

Richard Wood was no doubt a solicitor, as were many other clerks to such authorities and as pertains to-day. Richard was probably the father of John Wood 1765 - 1856, a solicitor of Melton whose son of the same name followed him in the family business. 'So Richard, whilst I have dug your name out of the dusty portfolios of the past, I must mention your son John, who was 'lived again' in memory at exhibitions held at Aldeburgh and in Ipswich in 1953. Your son was one of the "Six Suffolk characters" chosen to be "recorded in papers," as one who had contributed to the life of the county and who left records which others have considered worthy of care for posterity. And you too Richard, have achieved equal merit in the faithful accounts which you have left.'

Every record of the meetings of the old Commissioners of Sewers is preceded by the full preamble of the Statute, 'Knowye, for as much as the walls....' This, together with the minutes was hand written in a good neat flowing style. After good service, the clerk should have been able to quote the preamble word for word! At this meeting in April 1779 the members reported on a survey of the sea walls in the three parishes of their level and it was agreed ... "that a new wall be raised and made cross the Salts and creek from wall to wall within the level aforesaid, (part on this level and part near Bowman's bridge) and also to lay a sluice". Two other new walls were agreed, one of 60 rods and one of 100. Tenders were put out and [*Page* 131] at a subsequent meeting, the Commissioners agreed to accept the estimate of Mr. Balls, a carpenter of Middleton. The specifications included the building of a new wall and the cutting of a delph ditch. The wall was to be 9 foot high from base when settled, with a 30 foot base, 3 feet wide at the top and 20 ½ feet on the slope. The delph ditch to be 14 foot wide, 4 ½ feet deep and a 9 foot bottom width. Mr Balls' estimate for this work was £1 7s. 6d. per rod (5 ½ yards) and he agreed to find his own barrow and timber, for which provision he allowed £7 1s. 0d. on the satisfactory completion of the work. Hand labour, wheel barrows and 27 shillings and 6 pence per rod – that was in 1779. With the use of mechanical excavators and the present cost of labour and other incidentals, such an undertaking would cost about £24 per yard run!

---

5 [*Trist's footnote reads*] 1. Proceedings of Commission of Sewers for the level of Hollesley, Alderton and Bawdsey in Suffolk: 1779 Geo III.

In May 1779 the Commissioners levied a rate of 15s. per acre and a reference is made to the land of John Sheppard and Robert Vertue which on account of previous flooding, was not rated. As the walls had been repaired, it was decided that the rates were again payable. This was reasonable and more recently, the same has been agreed.

On the 15th May, Richard Wood the collector, records the balance of the commissioners at £645 15s. 0d. On August 25th the clerk enters, '1 horse for 5 days drawing stuff to the sluice 7s 6d.' And then writes a neat entry at the foot of the account: – 'By my salary on ye 2d. rate of £86. at 6d. in ye pound … £21 10s 6d.' In June 1780, the clerk recorded no difference in workers wages, '16 days work of labourers, 2s.' And at the foot of the account, he forgot his previous spelling and used the phonetic at that date … 'my sallery £21 10s. 6d.'

The Common at Bawdsey was let by the Commissioners to John Willett, Joe Hollowhead, John Brady, John Cooke, Robert Page, Jeremiah Pipe, David Ramsby, Robert Knight, Thomas Denny and John Clarke all yeomen of Bawdsey, for 5 years at £10 per year excluding rates. No doubt some of the descendants of these yeomen are still in the parish.

[*Page* 132] The feed on the walls and the salts was also let by this authority and in 1780 the Rev. Denny Cole appealed against the letting of 7 acres of saltings which he contended were his private property. The Commissioners appeared uncertain of their position and the clerk was instructed to obtain counsel's opinion. Subsequently Mr. Cole agreed to the letting and the 78 acres were advertised in the Ipswich Journal (now the East Anglian Daily Times). At this same meeting there was another case to settle. It appears that a watercourse known as Hollesley river was obstructed by not having free passage though a trunk laid across the river to the land of Sarah Williams and Mr. Whimper Bready. The commissioners ruled that the obstruction lay with Bready and that 'he should remove the nuisance or be proceeded against as the law directs': or as the Statute directs, "by fines, pains or other like measures after your good discretion"!

In 1796, the Rev. Richard Frank, D.D., took the chair and William Kemp the landlord of The Swan at Alderton was appointed bailiff. With meetings taking place at the Fox, Hollesley, and a publican landlord as a bailiff, the commissioners no doubt were able to look after their victuals when returning from their survey or retiring in Committee. Kemp had to 'execute all warrants, precepts and summonses' … relating to 'the Repairation, Amendment and Reformation of the Walls' … this is very nice 'Suffolk' and akin to the present day use of the word re-creation, with prefix emphasis for the word REcreation.

The records contain many references to the grazing[1] [*footnote in text*] of the salts and walls, [*Page* 133] always specified for sheep grazing only and there is little doubt that their treading helped consolidation and the close feeding of the grasses on the walls created a compact turf which was more suited to withstand tidal erosion.[6]

6 [*Trist's footnote reads*] 1. The East Anglian Daily Times of the 11th June 1953 reported a meeting of the Southern Area Land Drainage Committee of the East Suffolk and Norfolk River Board at Ipswich, which echoes the old practice of grazing the walls and saltings. Before the meeting, was a letter from the Clerk to the Walberswick Parish Council in which complaints were made that "the River Board paid insufficient attention to the knowledge of local men'. The letter also pointed out "that it was the rule that sheep grazed for so long each year on the river walls." "The sheep stamped the walls, thus keeping them firm, and closed the rat holes, which unclosed, weakened the walls and allowed the river water to percolate and scour into them".

This feature can be seen on any badly managed pasture which is under stocked. The tall and aggressive grasses get away and smaller species are crowded out by lack of light, whilst the dominant plants become single until the bottom of the sward becomes thin.

On the 14th May 1796 at the Alderton Swan, a matter relating to the letting of salts was settled by a number of jurors called before the Commission. They gave evidence that the saltings had long been grazed by the occupiers of land abutting them and this right had not been challenged. The jurors did not agree therefore, that the Commission had a right to charge for grazing on lands abutting; but it was agreed that the 21 acres purchased by the Commission from local abutting occupiers for the purpose of raising a new wall and delph, gives to the level, a claim on the Court of Commissioners to maintain the walls and cleanse the ditches, by reason of frontage ownership. The Commission therefore, have a right to let and charge for the grazings on their walls and saltings.

Here is further witness of the justice meted out and the democratic agreement by arbitration of a jury – with both sides recognising their rights by custom and their obligations at law.

Over the period 1779–1806 of these records, there are constant references to the letting of sheep grazings on the walls and saltings; and there is no doubt that the close grazing and the finer grasses created a tight turf, whilst the coarse long grass such as the sea couch would act as some protection to the bank, if there was an overspill of tidal water. [*Page* 134] At that time; the 21 acres of saltings belonging to the Commission were let for £10 11s. 6d. per year for a term of 10 years, a high rent for such poor feed as could be found on a salting.

The period of about 25 years at the turn of the eighteenth century was an important era in the annals of agriculture, for real improvements were taking place in farming technique. By the middle of the 18th century, Lord Townshend had set a new pace in cultivations on the land. He introduced the root crop into the rotation and earned the name of Turnip Townshend. In the last quarter of the century, Coke of Olkham interested himself in agricultural education by inviting farmers and workers to his sheep shearings, and by his practical demonstrations and discussions he can be regarded as one of the forerunners of discussion groups and farm walks which over the past fifteen years have been revived to the great advantage of many who seek to learn in farming. Then Robert Bakewell of Dishley looked with an indignant eye on the standard of our livestock and their production; and following the example of improvements made in crop rotation, he selected stock and produced animals worthy of the name. The revolution was underway, and more pioneers of thought and action sprang into the limelight with new ideas for improvement: we see Joseph Elkington of Warwickshire trying out new methods for leading off spring water; Elkington's thesis on waterlogged land is true as it ever was and in our plight to-day for increased food production, good crops by modern standards are even more dependant on well drained land.

[*Page* 135] At the turn of the century Arthur Young surveyed Suffolk on horseback. As the first Secretary to the Board of Agriculture he presented a report[1] [*footnote in text*] in 1794 which was later circulated to invite comment from the public of the county.[7] In a series of critical observations, Young records the type of cow

---

7 [*Trist's footnote reads*] 1. General view of the agriculture of the County of Suffolk with observations on the means of its improvement. Arthur Young 1794.

found in the county – 'the breed is universally polled … a clean throat with little dewlap, a snake head, clean thin legs and short: a springing rib and large carcass; a flat loin, the hip bones to lie square and even; the tail to rise high from the rump'. Well, how does his description tally with the modern Red Poll? Young says that the quantity of milk recorded is considerable for all good herds have some cows giving 8 gallons per day in June; and it is common for cows to average 5 gallons per day.

I can find no notes of his relating to marshland, but he leaves a caustic comment on other pastures …. 'the management of meadows and upland pasture in this county in general, can scarcely be worse' …! But let us hasten to add that his praise for the standard of husbandry in Suffolk 'is high and esteemed in enterprise'.

In 1810, we see the introduction of a Land Drainage Act for Minsmere which was given Royal Assent on May 18th and out of this Act, Minsmere saw improvements. On an old map of 1813 in the possession of the late Mr. F.E. Holland of Leiston Old Abbey who died in 1953 at the age of 93, the 'Old Sluice' is marked at Coney Hills below Minsmere Cliffs. The New Cut, the large high level dyke running through the centre of the level is shown cut to the wall, but no sluice is marked in its present position.

[*Page* 136] Whilst at Minsmere, let us look at some interesting correspondence[1] [*footnote in text*] which took place a hundred years later in 1906.[8] The correspondence refers to the 'bent hills' – the sand dunes planted with marram grass between the sea wall and the beach – and stresses their value in protecting the wall by breaking the full weight of the tide. It appears that in cases where the wall had been broken, it occurred where the sea had eroded the benthills; so some protection by groyning was necessary to protect the benthills. At this time it was said that 'a groyne was erected near the north end of the wall many years ago but the sea washed it away?' A second groyne is recorded halfway between the sluice and Dunwich and which in 1906 had stood 8 years and was then expected to be washed away. The discourse then turned to faggotting and referred to lines of this protection which were still to be seen in the sand, having been put there thirty years previous, but which had been so neglected that in 1897 there was a large breach in the wall. The report records that faggotting in three lines two yards apart was erected and the marram grass was planted to hold the sand.

This vulnerable corner of Minsmere level has been a nightmare for the engineers for many a long day. The old Catchment Board were proposing to sponsor a scheme for the Internal Drainage Board in 1950, for the erection of several groynes in this area of the Coney Hill wall, but the plan has not yet developed. In this corner of the level up to 1815 the cost of defence works had been £12,000. From 1815 to 1906, expenditure in normal years was about £250 per year, but the report records in lamentation that the abnormal years occurred rather too often! Here are some of them as examples: 1837, £494; 1847, £1402: 1848, £746, 1859, £402; 1864, £700; 1875, £800; 1882, £1714 and 1899, £830.

Concerning the Iken level, there is more interesting information beautifully recorded by hand for the period 1871–78. Sitting at the Crown and Castle at Orford in 1874, the Commissioners reported a very happy situation … 'The estimated costs [*Page* 137] of work to be done on this level (in 1871) with certain annual expenses

8 [*Trist's footnote reads*] 1. A report from the Hon. John de Grey to the Clerk of the County Council at Ipswich.

is about £51 10s.'[9] [*footnote in text*] The account shows a balance in hand of £15 1s 3d. and we recommend that a rate of 1s. 6d. in the £ be collected, which will raise the sum of £51 13s. 0d. and a balance for any casualty.' It is clear that their maintenance work was well in hand, bit it also gives a clear picture of the change in the cost of the works!

The minutes are valuable in their detail and clearly show that the information was obtained by members' inspections. There were laborious introductions … 'In the fortieth year of the reign of our Sovereign Lady … and present, Rev. John Maynard, Chairman … the scribe records 'Messieurs James Chaplin, George Rope …' Does the present Board record the presence of Monsieur James Mann?

All reports made, had to receive attention forthwith and detailed arrangements were carefully recorded.

> At Bognay sluice, the stops to the door require to be made fast
> At Stannay Pt. about 5½ rods require to be filled in
> At Bognay ploughed marsh 6 rods want siding and about 3 loads of 4 foot pile will be required

On 31st January 1877, there was a high tide which overtopped the wall at Iken Church.

'At Isles Ploughed marsh, water over and requires topping and at Mr. Coopers Reedling and Bognay ploughed marshes, about 100 rods require topping'

At Bognay point there was a small scour requiring attention and the sluice was reported as being in a bad way.

'Lantern marshes, Havergate Island and Mr. Pettit's marshes still under water on July 14th'.

[*Page* 138] The 1s. 6d. rate of 1874 had risen to 4 shillings in the £ by 1882 when 'the estimated cost of work to be recommended is £140 11s. 6d.' From the meeting of July 12th, 1878 there is a detailed record of a considerable amount of work, for an extraordinary cost.

> The Ferry Wall, about 100 yards from the new work towards Iken requires topping
> The wall delph from Mr. Chambers sheep bridge to Orford Quay requires to be cut and cleared
> Hall Hill watercourse requires to be bottomfayed
> Ox watercourse requires to be cut and cleared
> The piling at Ox sluice run requires to be repaired. All the new works on this level required to be sludged.

This work was estimated to cost £156 0s. 7d.! In the same minute, there is a sentence which frequently occurs throughout the record … 'destroy all rats and stop holes throughout the level'!

The meeting of September 11th, 1899 records the heavy burden created by the disastrous high tide of November 1897, when the Rev. Beaufry James St. Patrick, Chairman asked the meeting to pass the accounts for the work completed:

9 Trist does not provide any corresponding text for this footnote.

Chillesford Lodge walls	£355	2s.	6d.
Gedgrave Walls	£1121.	12s.	3d.
Lantern Marsh Walls	£1090.	19s.	4d.
	£2547	14s.	1d.

The Commissioners recommended 'that a £2 rate be collected which will raise £2825 and have a balance in hand of £278 5s. 11d. to carry out the work of topping and sludging the walls which have already been commenced.

To the north east of this level, across the river, lies Aldeburgh and Slaughden, both of which can contribute a great deal to the history of the Suffolk coast. The sea defences at the south end of Aldeburgh have involved the expenditure of many [*Page* 139] thousands of pounds, and the problem is not yet solved, in spite of extensive groyning to hold the shingle in the narrowing neck of Sudbourne beach ~~Orford Ness~~.

In 1872 the Commissioners for the Iken and Orford level were perturbed by a Bill under consideration for the construction of a harbour at Aldeburgh, by cutting through the shingle bank and petitioned the House as follows:

'In the House of Commons'

In the matter of a Bill for confirming the certain provisional orders made by the Board of Trade under the General Pier and Harbour Act 1861 relating to Aldeburgh and Lynmouth.

Petition,
of the Commissioners of Sewers for the level of Iken and other Levels against, the construction of a harbour at Aldeburgh as proposed.

'That by order of the Board of Trade to Aldeburgh proposed to be confirmed by the Bill now before your Honourable House it is sought to sanction a Cutting through the Sea Beach at Slaughden Quay near the town of Aldeburgh'.

The petition continues, that if this is done, the river walls will not withstand "the force of the waters" – the land will be liable to flooding and the rates to be levied for repairs will be a severe tax on the rate payers. The petitioners considered that the mouth of the river at Orford Haven would become choked up and impassable for vessels, that property values would fall and trade would be injured. It further contended that Aldeburgh could be approached by river and the proposed harbour would not provide such good shelter as at Orford Haven. The petitioners finally gave their opinion that the resources of the promoters were inadequate for both the scheme and [*Page* 140] for meeting any compensation to land owners who might suffer by flooding. It was also the Commissioner's opinion that a bar of shingle would develop across the mouth of the harbour.

Well it never happened and no doubt the petition was well considered: but as the years go by the scene changes here as elsewhere up and down the coast. Forty years ago, a small boy who from an early age loved the marshes, used to walk along the shingle to the Lantern Marshes and lay there in the sun to sleep or watch the birds. At that time, the width of the shingle between the Martello Tower and the Yacht Club was about 100 yards at high tide. To-day[1] [*footnote in text*], it's about

40 yards.[10] And that small boy is now Anthony Hurren, farmer of Park Gate Farm, Stratford St. Andrew. He still loves the marsh, but in recent years he left a farm with a marsh where the river wall almost broke his heart.

Anthony Hurren comes into the picture at this point, because he has a theory which is contrary to the Commissioner's position. In a conversation with him sometime after the 1953 flood disaster, he brought out the idea of breaching the shingle at Slaughden in a south easterly cut, so that the tide would pass into the Alde and in time build up a shingle bar on the north side and thus protect the vital area at the south end of the town of Aldeburgh.

I cannot and would not comment on the wisdom of this idea – anymore than I would give an opinion on the future of Snape bridge! But Mr. Hurren has lived 'hereabouts' with a very observant eye for sufficient years to assess the repercussion of the tides on the banks of the Alde.

Before we leave this problem point, let me quote from someone who signed himself 'Longshoreman', … "where Davey Jones has been once, he will come again".

For the level of Falkenham and Felixstowe, I did not find much in my researches. There is an interesting hand written record on vellum dated 1832 [*Page* 141] which records 'a rate and assessment of four shillings an acre, ordered to be made on the owners of lands within the said level … for and towards repairing ….' In 1845, the rate was 2s. 6d. per acre and 1s. 6d. in 1881.

A few years prior to 1875, the private enterprise of Col. Tomline in building a new wall caused a stir in the Commissioner's camp. The Commissioners for the level had previously constructed a sea wall from a point called East End, along Felixstowe Common and the Colonel had raised a new wall, the end of which encroached on the delph of the Commissioner's wall. Subsequently an agreement was drawn up in 1875 between 'The Commissioners of Sewers for the Falkenham and Felixstowe level, John Chevallier Cobbold of Ipswich, George Josselyn of Straford St. Mary, John Patteson Cobbold M.P. of Ipswich, Charles Thomas Cooper of Henley for the Commissioners and Col. George Tomeline of Orwell Park, Nacton'.

The commissioners sought to prove that at the north of the intersection of the new wall with the old wall, a weight of water would be thrown on to the old wall and Col. Tomline should be answerable for any tidal damage to the Commissioner's wall, but the Colonel did not agree.

At the price of some fencing and one shilling per annum, the parties came to an agreement whereby it was acknowledged "that the Commissioners were at the time of the building up the new wall, in possession of the wall slope footing foreland and ditch over which the new wall is constructed". Col. Tomline agreed to pay one shilling on January 1st each year as an acknowledgment of the title to make use of the footing; and also to fence off the Common from the new dry part of the Commissioner's ditch. We can picture this great landowner, whose estate covered several thousand of acres between the rivers Orwell and Deben, [*Page* 142] solemnly sitting sown on the last day of each year and writing 'Pay the Comissioners of Sewers – one shilling only'.

The method of repairing and the general maintenance of sea and tidal river walls remained much the same up to comparatively modern times. It is only in the last fifteen years that we have seen the introduction of excavators, bulldozers and other

10 [Trist's footnote reads] 1. Feb 1953.

powerful mechanical devices for moving materials: and it is only over the same period that anything faster and more powerful than a horse and cart or sledge was used for transport on the marshes. As speed and power has come to our aid and modern times have changed the standard of living conditions so the cost of this work has increased.

What did it cost and what sort of a pay packet did those men bring home who used to toil in all weathers, both day and night?

To keep strict accounts is a bind and often a shock when the books are balanced! What can an old account book arouse? Many of these books still found on farms, give invaluable history of cropping, of expenditure and income and weather records.

Over the period 1845–1880, there are four books[1] [*footnote in text*] of "Marshmens work" (in Suffolk) which give an interesting record.[11] The earliest copy was kept by Thomas Freeman in good handwriting with accurate entries and very sensible phonetic spelling of marsh vocabulary which even to-day is more often spoken than written. In February 1845, Aldred, Hurren, Fiske, Sowyer, Hatton and other Suffolk names were 'making good the Blythborough wall at Bulcamp Farm' – how these men must have turned in their graves in the past twenty years, whilst the tide has daily flowed over the Bulcamp marshes! The foreman had a guinea a week, most of the men received 12 shillings, some 6s. and some 4s. and boys 1s. 4d. per week. This was standard of those times, paid by bush telegraph agreement for there were no Wages Boards such as we have to-day.

[*Page* 143] In 1834, eleven years before our diary opens, a new Poor Law Act was passed. Up to this time, the old Act of Elizabeth I had its own way of looking after the poor by the responsibility being laid upon the parish to succour its own poor and needy. The enclosures of the nineteenth century brought poverty to the farm worker and he also suffered with the high prices of food caused by war. The new law of 1834 destroyed the old parish system of help for the poor and made a number of parishes a 'union' with one institution, which was later given the name of 'workhouse'. Here the poor found themselves under officialdom with rules contradictory to Christian ethics. This was part of the framework of agricultural society which lasted throughout the century in which England becomes prosperous. However, such was the law and the attitude towards the worker, who has now risen to a place worthy of his contribution to agriculture.

But back to our diary and let me quote rural English which gives apt description of work on the marshes. In the early summer of 1845 'Bottom fighing wall delf in Mr. Farrow's scant marsh and on Mr. Freeman's marsh joining the Walser crick': 'the foreman 10s. 6d. and men 6s. per week.'

Bottom fighing is an expression still used in Suffolk and spelt the same way. It is cleaning out of silt at the bottom of the ditch, which has been borne by the winter rain water off the land and the sides of the ditch. A scant marsh is the marsh adjacent to the river wall and the delf (or delph) ditch; the latter is usually wider than the ordinary run of marsh ditch and runs along the foot of the wall known as the folding or basin, dividing the scant marsh from the river wall. In Norfolk, and in north Suffolk, the delph is more frequently called the soc dyke.

[11] [*Trist's footnote reads*] 1. From records deposited by the Rt. Hon. the Earl of Stradbroke from the Henham Estate.

Then 'Youse of 2 barrows 2 days … 0/4d.' The barrow remained the chief transport and dumper on the wall for almost another hundred years and not yet disappeared – but it is no longer rated at a penny a day! [*Page* 144]

Making good the sluice at the new wall	2s.
Crafting of earth and making of dams at Walser bridge	12s.
Filling up the dike in Mr. Freeman's marsh	7s.
Making good drinking places in the fooles waterin marshes	7s.
Bottom fighing and leveling at Farrows	6s. 6d.
Turning up the foot of bottom fighing and making good drinking places in the fooles warterin marshes, 6 men employed at per week	2s. 2d.
Cutting and drawing the river and dikes in the Choesey marshes	2s.
Bottom fighing the reedroad dike	[*blank*]
Showling and depining Walser crick	13s. 4d.

'Crafting of earth' probably means the conveying of clay from some point to the river wall for repairs or 'making good' to the banks. At this period, when both wages and the time factor were so different compared with conditions to-day, good strong clay was selected and borne by barges up to the wall, hence the word 'crafting' or conveyance by craft. Drinking places on the marsh are cut out inclines to the ditch so that cattle can drink without treading down the ditch bank or falling into a deep ditch. The are usually fenced off at the sides and often wired across the centre of the ditch, to prevent cattle crossing to other marshes or trampling the ditch. The 'fooles waterin' marshes still bear the name and lie to the north of the Wolsey Creek and on either side of the Halesworth–Southwold road. The origin of their name appears to be lost. 'Cutting and drawing dykes', is an expression still in use and something which still has to be done. It is the cutting of reed and other growth on the banks and in the bed of the ditch. If this is not done, the annual growth dies and topples into the ditch as debris and impairs the smooth flow of the water.

[*Page* 145] The reedroad or rond is sometimes an immature unreclaimed marsh inside the wall or more often a riverside salting. It provides salty rough grazing and is unprotected by the tide except for a small low bank and is more often a high salting only covered at the spring tides. The word reedrond would indicate this position, because the common reed (Phragmites communis) whilst being a highly salt tolerant plant, is not so able to withstand constant inundation by the tide. The sea arrow grass (Triglochin maritima) a common plant on these salts, is frequently known as rond grass. On the other hand, the 'reedrond dike' can also refer to a ditch cut through a 'fleet' in a marsh. A 'fleet', examples of which can be seen in the Queen's Fleet at Bawdsey and the King's Fleet at Holm Hill near Felixstowe, is an old river bed and the area on either side of the ditch (the surviving course of the river) still remains a soft reed covered bog over the area of the width of the old bed of the river.

"Showling and depining Walser crick" … this is a larger job than bottom fying (modern spelling). Showling would entail the removal of silt, sand and gravel. 'Crick' is the Suffolk phonetic for creek, as the record of Walser crick undoubtedly refers to Wolsey's Creek.

In September 1848 one man was 'making good the walls adjoining Mr. Girling's marshes … 10s. 6d.' Mr Geoffrey Girling of Reydon Hall who farms 500 acres on the north bank of the Blyth, is probably a relation. His family has been at the Hall since 1897 and prior to that date they farmed the Whitehouse at Frostenden for 200 years. Over the years he has had cause to see the men 'make good the walls' and even find sluices which were not known to exist! Then Harry was 'cleaning out the sheep wash' and took 2 shillings for his work and no doubt earned every penny of it! Whilst another was 'driving piles to make Mr. Freeman's wall joining Walser crick' and his pay packet contained 1s. 10d. for the week!

'After the rats on the walls' ... for 3s. 8d., is a frequent entry [*Page* 146] and was considered a most responsible and important job. The entry is expressive and the words "seein' after" are still in common use.

From 1850 a new scribe takes over the book of marshmen's work and does not seem to concern himself with so much detail. In 1853, one man has 12 shillings for the week, whilst daymen receive 4s. for work on Mr. Girling's marshes and 'crafting of earth to make good Mr. Crow's wall below the Union'. An entry in December for 'slubing the new wall ... 5s.,' is followed by 'after Do and Do ... 15s.,' whether the recorder meant the work was getting clay for facing the wall or whether he was 'after something or other', but had forgotten what ... I do not know.

The next entry has an interesting word. 'Shroughing up earth at the back of the new wall': this word has the same meaning as 'slubing' and means the plastering of wet mud which was dug out of the saltings near the water's edge and barrowed back to the wall over planks. It was then placed on top of the wall and on the sides and firmly patted down with the back of a wooden shovel. This was not a wall building operation but one of maintenance. Where a wall developed cracks in the summer or had surface damage by cattle treading or where repairs had been made and new cut turves had been laid, the slubing spread wet mud into the crevices and sealed the surface.

When F.J. Freeman took over the books in 1870 he recorded in a copper plate flourish with surnames preceding the christian names; but as a record, it lacks the detail of the early books. Wages show little or no improvement for 'cutting and drawing the wall delf' 2s. 6d. was paid: bottom fying was paid the same, with 1 shilling extra for the foreman. A gang working at this time comprised Robert Croford, William Cobb, William Saunders, George Cobb and Richard Jones: perhaps some of their relations still work on the walls of the Blyth river? How could they manage on 16 shillings for a six day week driving piles, and would they fancy 18s. 8d. for seven days reed cutting?

[*Page* 147] By 1880, Richard Cook and George Stockdale had joined what appears to have been a permanent wall gang and they replaced Cobb and Jones: And so the bottom fying and slubbing continued and it still has to be done. One last item is worthy to relate from these books and is entered on October 1883. 'Crawford and 3 others, 10 tides 2 nights, 60 hours ... £4 2s. 0d.' Their reward is one thing, but 2 nights is significant of the urgency in working with the state of the tide. If a job had to be done, then all the hours allowed by the tide had to be worked. With a moon, their task was made easier but otherwise the dim light of a lantern would not seem much comfort or help!

For almost 350 years, the Statute of Sewers of 1532 had been the law governing the state of repair to our sea defences and British agriculture is greatly endebted to the Commissioners who assiduously carried out their allotted duties.

From 1853 to 1876 our farming improved and prospered whilst the countries of Europe finished one war and then started another. Prices rose high and when profits soared, the same competition for farms became apparent that we see to-day in the rents which some were prepared to offer. By 1879 there was a sudden fall in prices. It was an exceptionally wet season with rain from harvest up to the New Year and river fluke took toll of millions of sheep. Many farmers were ruined and could not pay their rents or their dues to the Commissioners of Sewers who in their wisdom refrained from prosecution.

From this time up to the turn of the century, circumstances gradually prevented the Commissioners from exercising their functions through lack of funds, and their 'commission' slowly dwindled until they ceased to function. But the work had to go on and landowners took over the responsibility of maintaining the walls defending their own properties.

[*Page* 148] The men of the walls or lengthsmen as they were generally known, were marsh men who lived literally close to the water as the farmworker lives to the land. They were men who 'looked' at a sea wall as a hoer would 'look' as his sugar beet. They knew the vulnerable places in their length and where good plugging material could be found: they knew the wiles of wind and tide, and when they were likely to be wanted.

Very few of this old school now remain. Those who are still alive are not unnaturally critical of some modern methods of maintenance and repair work. In their time, it was all work by hand and to-day, machinery supercedes the shovel and the barrow: but each age has its criticism of the new order of the time and each has its own argument.

One member of this old corps of wallmen lives in Waldringfield on the Deben, where he has worked in the vicinity all his life. Mr. William Bear worked for over forty years on the river walls between Martlesham Creek and Waldringield. In 1944 he came 'off the wall' and took up casual ditching work for the Catchment Board and local farmers. Whilst on the river wall, he worked for the Prettyman estate, then responsible for the walls and later for the Board who subsequently assumed responsibility.

Some weeks after the 1953 floods, I was talking to Mr. Bear over a pint of beer in the Maybush at Waldringfield and our conversation quickly drifted on to the river walls and the repair work proceeding under the direction of the River Board.[12] 'How are they goin' alon' … I spose its all this machine work now?' 'Yes', I said, 'the job is too big for hand labour and excavators are now the order of the day.' He thought a little and then said, 'I new'd it wuss a big storm, but they dunt repair'em an' look arter the walls as they should'!

I agreed the importance of maintenance work and the constant eye of small things which grew into trouble, such as rat holes. This drew my [*Page* 149] friend …, 'You talk o' rats … dew you find a rat hole in the bank, you muss dig'un out to is full length, fill'un in and rammun with mud took from the land side o'the wall … Thass the right way, but I've see'd em dig in about a fut, ram with earth and put an' owd rag stun uvver the hole!'

Illustration and experience was retold whilst I listened intently: and I also thought of the practices of the past which in our modern age are being lost to the cause of speed and the paramount demagogue of economics.

I was anxious to know when the river wall was breached in Martlesham Creek against Sink and Sluice farms. Mr. Bear could not exactly remember the date, but it appears to have been sometime before 1920. At any rate, he had a vivid recollection of the state of the wall about this time for he remembered pushing his spade down to the handle through a crack in the top of the wall! But it was another story of rat holes which he wanted to tell me and which I relate.

[12] Maybush Inn, Waldringfield, a short walk from Trist's home.

'I mind the time I wuss workin' on the wall at Martle'shum crick, that wuss before the wall brook. Sudden, I hear the foreman … come yew over here, he say. I go alon' an he point to a rat hole I'd left on the inside o' the bank. This 'ere ont dew, he say … Now I wuss working on the outside o' the wall … I stand on the top o'the wall an' jump an' made the wall rock. There yew are, I say, its the owd footins … we can't goo arter they rats now, dew we'll have a breach …. we muss work on the fut o'the face o' the wall and see arter rat holes arterwards …. Yew'd ha thought he'd a known better.' There you are, the importance recognised, but first things first. There appears to have been no breach of the peace between Mr. Bear and the foreman in charge of the job, but it was timely that the owner of the property showed up soon after this incident and heard something of the details. 'Leave him goo on his way, dew he ont work with us' … Mr. Bear's wisdom prevailed and he worked on – the right way!

[*Page* 150] Incidents of wind and tide are not uncommon on the east coast. Sometimes the damage to both wall and land is comparatively slight and sometimes it is more serious: but to someone it is always serious! A deep fresh water flood may retard land work for a year, but salt water flooding leaves behind a far greater problem.

Of some things which do not directly affect us, our memories are very short: and even those who are near at hand, find that their memories fail for detail. Of those affected, I find their memories good – who would fail to remember the year when a herd of cows was left with ten out of fifty acres of summer grazing? So let me record some of the flood incidents on the Suffolk coast in the past forty years.

From November to March, those who knew of the combination of factors which produce a dangerous tide, keep vigil. A full moon, a high spring tide and nor-west gale in unison, are ill omens!

In 1915 it blew hard on the Breydon Water in the north of the county and the tide came over the wall and flooded about 400 acres of the Caldecott and Belton marshes. With the devils combination set for foul, these marshes lie in the 'right' place on the north east corner of the river bank. In 1921 there was a break in the Blyth river wall near the Bulcamp marshes and in the following year, the same place collapsed again with a further two breaks near by. The work was left and the flooded marshes abandoned. Year by year, the breaches were eroded by an increasing volume of tidal water, until the banks of the river near Blythborough were shattered in their abandonment. Soon after the original events, the County Council were obliged to erect an embankment to prevent flooding on the main road.

[*Page* 151] To-day, the Angel, the Bulcamp and the Blythburgh marshes containing 486 acres are still flooded on every tide. In 1945 a scheme was promoted – it is still being promoted or at least it lies tepid. It is a long story of the future of the Southwold harbour, river amenities, town sewers and the land! Maybe one day the Blyth will again flow in its proper course and the marsh again become green.

In 1924 and again in 1935 there was a furious sea raging in the Alde which sent its fury up to Snape. The marshes at Dunningworth Hall were breached and doubtless land within the Ore was flooded at the same time. In 1938 there was one break in the wall at Iken and the water was on the marshes for three months. Between the railway and Snape bridge there was a break in the bank which caused extensive flooding through the buildings of Snape maltings and the Dunningworth wall broke and once again the 108 acres went under salt water. At Minsmere, the sea battered its way over the shingle below the cliffs and made a gaping hole in the Coney hills wall, to flood the level north of the internal bank above the old Minsmere river. At the same time, the gale raged on the Deben and breached the bank on the Hemley Hall and flooded 90 acres of its marsh together with 30 acres of marsh of the adjoining farm. This break was not only the cause of lost land, but of long drawn out litigation between the owner of the land and the Catchment Board: and a case which was followed with considerable interest, for it was won and then lost and very nearly went to appeal in the Lords.

In 1939, yet another tide battered at the Dunningworth Hall marsh wall and breached it again. On this occasion it took the bottom of the wall out for about ten yards and the marsh was flooded for over two months: this was the fourth flood in fourteen years. When Mr. Anthony Hurren took over the farm in 1941, it was not surprising that the marsh was a mass of creeping thistles and sea aster.

[*Page* 152] The owner, Lord Ullswater and new tenant decided to go ahead with the slow process of reclamation, in spite of the many set backs. The ditches were cleaned out and the fences and gateways restored. In the early summer of 1941, the landlord was served with an order by the Agricultural Executive Committee to deal with the thistles on the marsh! Mr. Hurren put three men on the job who scummed the marshes with scythes. In the same year, he had a present from his uncle in a quantity of three year old wild white clover seed which had not found a market. With no cultivations at all, Mr. Hurren broadcast the clover on about 80 acres. Mr. Hurren told me that germination of the seed was very slow and irregular. The plants which survived, remained shy of growth for a long time, but in four years he had a mat of growth. This account of white clove behaviour on salt flooded land is not surprising, but it is interesting that memory records that the best 'take' of seeds were on the two marshes (O.S 118 and 130) opposite the 1939 breach.

In the same year came war and with all its attendant horrors, came ruin for many hundreds of acres of marsh and arable land in Suffolk. Our shores had to be defenced against one Hitler and so valleys which presented a potential landing place for invasion, were inundated: in addition, thousands of acres were sterilized for battle training areas.

In the north, Kessingland level was allowed to flood with fresh water and 800 acres were taken off the list of grazing grounds. This unhappy level! Even before the war, it had its water problems in a sluice which was constantly blocked by shingle. After 1945, the marshes remained a bog, flooded for the best part of each winter until 1949 when the Agricultural Executive Committee excavated the ditches. The water then had somewhere to run, but all was not happy at the Benacre sluice. In 1950 a pumping station was installed with one pump, followed in 1952 by more power and an overhead concrete flume which conveyed the water from the Hundred river across the [*Page* 153] shingle. But the pumps occasionally broke down, the shingle filled up the exit to the old sluice and the river silted up on its bends and nothing seemed to go right for the future of the land until the late summer of 1952 when all was working well. The water had never been lower in the ditches and the land was at last drying. Next year, it was thought that the time had come when a safe start could be made on reclaiming the marshes from twelve years of idleness and desolation. For the past three years, several occupiers had made serious efforts in preparing the land for the plough by continuous mowing of the tall weeds – and then came another disaster, the great tide of 1953 which we will discuss later.[13]

[13] Evidence that Trist originally intended this chapter to precede a chapter on the 1953 floods in a longer work.

In the Iken–Sudbourne area, nine thousand acres were taken over as a battle school early in the war. About half was marshland and half was under arable cultivation. For seven years, this land lay wracked in the agony of neglect; the hedges grew high, its ditches became choked, its buildings and houses fell into decay. On the marshes, the grass grew, seeded and toppled over to join the debris of former years. And whilst nature ran riot, tanks roared over its plains and shells pitted the ground to train men for further destruction.

In 1947, this was a desolation of no-man's-land: the arable land was ablaze with golden ragwort in a thick bed of couch grass. The marshes were almost a bog through filled ditches. Subsequently a fleet of twelve ditch excavators restored a hundred miles of ditches, whilst many tractors hummed over hundreds of acres in full fallow.

Eventually the houses were restored, even to digging the gardens and restoring the garden gate; the farm buildings were repaired and then the roads were restored. Whilst the marshes were ditched with new culverts and new gates, they were only given lime and surface cultivation and it was the former occupiers who returned to their land together with some new entrants, who really restored the marshes to respectability. It was Mann and Kidner, Fulcher and Black and others who returned [*Page* 154] the plough to this land: who with the new tackle of heavy deep digging plough, rotovator and bulldozer, put a face on these marshes which had never before been seen.[14]

Whilst this area was under requisition, the sea again roared in the Alde in 1944 and the river wall was breached on the Hill and Hall farms at Iken. At the same time, there was another break at Dunningworth Hall and the marsh was again flooded, but only for two days. This time, they were prepared.

When Mr. Hurren took over this farm in 1941, it was agreed with Lord Ullswater who had assumed all further liability for the river wall, that the farm should have emergency stores in the shape of barrows, stakes and baulks of timber. In 1941 the long baulks were laid in across the break one on top of another, whilst the barrows tipped the clay outside to lie against the timber until it was built up to wall height. This was the first aid repair whilst the water ran out of the sluice. As soon as the tides lowered, the baulks were removed and the centre and back of the wall break was made up: and the stores then returned to their place with a new supply of stakes in readiness for another emergency.

The 1944 high tide also did damage on the Deben and as is to be expected, it hit at the weak spots. At Sutton Hoo immediately opposite Woodbridge, a break in the wall which had been repaired a few years before the war, again collapsed and flooded about 20 acres of marsh. Requests were made … but the tide still flows in each day through two breaches which steadily grow in width.

Farther down on the opposite bank a small break occurred on the wall defending the 35 acres of marsh which go with Cross and Hill Farms, Martlesham and a few acres belonging to Howe farm. Soon after the break, the wall was repaired but in a month's time another high tide burst through the repair and [*Page* 155] it was abandoned. To-day, the tide flows through at will and on the ebb it pours through the gap with an increasing scour under the foot of wall, as it rushes into a saltings creek on

14 Farmers at High House Farm, Bawdsey, Firs Farm, Sudbourne, Poplar Farm, Iken, and Crag Farm, Sudbourne, respectively (MAF 157/23, County Agricultural Relief Committee, Correspondence with Lord Mayor's National Flood and Tempest Distress Fund).

its return to the river. It is a length of wall well known to all who sail the Deben and to many others who lie under its lee with a gun patiently waiting for duck.

Then another five years elapsed and on March 1st, 1944 between 1 and 2 pm a nor-west gale and a high spring tide contrived to do their damnest once again. At Hollesley the sea poured over the shingle and backed up Barthorps Creek to swamp the Oxley marshes facing the sea: up the creek, it ran over the road and flooded the marshes of the Borstal Colony and its neighbours. The pumping station was surrounded by water and temporarily put out of action. On the Deben, the tide found several weak places in the wall. The little eleven acre marsh under the hill on Methersgate Hall was flooded from the breaches. The land is still flooded and the marshes abandoned, for the Catchment Board in considering their meagre resources had to decide that the cost of the repairs did not favourably relate to the value of the land. A little to the south by the sluice, there were two smaller breaks which were repaired to contain another small marsh. At Pettistree Hall still occupied by Mr. Tom Miller, 40 acres of marsh were under salt water for six days from four large breaches. The sluice continued to discharge water but was assisted in the final 48 hours by two pumps which were set up at the south end of the wall. The first aid repairs of the river wall were carried out under the supervision of the Catchment Board, by European Voluntary Workers employed by the County Agricultural Committee, the E.V.W's, who collectively represented many central European countries.

Further down the river, there was a break in the wall defending the south bank of Kirton Creek. On the opposite bank, the wall defending the little marsh below Keeper's cottage and Ramsholt church was breached in three places. [*Page* 156] This is a marsh in miniature; surrounded by a sandy hill crowned with trees, on which stands the church with its charming tower, which smiles a rosy pink in the afternoon sun. All those who sail up and down the Deben or walk its banks know Ramsholt and 'its Arms', set under the hill and just up from the muddy shore which once boasted of a quay – now a mere skeleton of a landing stage.

On the other side of the peninsular on the left bank of the Orwell, the bank of the Stratton Hall marshes was breached and about 120 acres were flooded – and the tide still flows in at will, whilst the walls gradually fall into decay.

On the Orwell, the river swept over the Strand to flood a few acres of arable on Wherstead Hall and further downstream, the tide came over the wall at Hill House farm to cause a setback to good pastures in the occupation of Mr. David Wrinch.

On the River Alde, the Dunningworth Hall marshes were flooded for the sixth time since 1924. Fortunately the breaching was not large, about 12 feet wide and took off about 4 feet of the top of the wall, which was repaired by the estate some weeks later: but the tide swirled around with a great force under Iken cliff and seriously cracked the wall for nearly 300 yards.

Further up the coast, the sea came over the shingle banks and flooded many hundreds of acres of marshes. Between Dunwich and Walberswick it broke through to flood the Reedland and the Dingle marshes: it covered the Town and the Corporation marshes and poured through the wartime breach in the Point marsh wall and into the Westwood marshes right up to the head of the valley, covering a total of over 1,000 acres. At Easton broad, the tide swept up the valley over the road at Potters bridge and halted below Jays farm. At Covehithe, it overtopped the broad and flooded the Poors marsh and also at Benacre where the shingle facing the broad affords no protection from a raging sea.

[*Page* 157] And so the eternal battle between the sea and its lost ground continues. As the years go by, we see an ever increasing mechanisation in the repairs of the damage. Small and large excavators, some with giant jibs, dumpers and bulldozers of many sizes. Every four or five years, then perhaps a breathing space of ten years and the wind ruffled tide will batter at the defences of the marsh.

Now let me turn back the pages of history again and trace the story of the River Blythe over the past 200 years. It is a strange history as far as my own interests are concerned for it concerns the improvement of a river for commerce and damns agricultural enterprise. The Commissioners were not set up under the Act of Sewers and had a different function, but here is a story of the passing of a navigable river complete with the 'death pangs' of the Commissioners recorded in the minutes. 8th June 1757. 'At the first general meeting of the Commissioners appointed for putting in Execution an Act, passed the last session of Parliament (Entitled an Act for making the River Blyth navigable, from Halesworth Bridge in the County of Suffolk, into the Haven of Southwold) at the House known by the sign of the Angel in Halesworth aforesaid, the Eight day of June in the year of our Lord, One thousand and seven hundred and Fifty seven'.

So runs the first entry in a large vellum bound volume which contains the minutes of this Commission from 1757 to 1839. There were 18 members present including Sir John Rouse Bart., and the Rev. Ralph Blois.

As previously mentioned, the Blyth had been navigable for many hundreds of years, but the arrival of a commission to take things in hand was not only a sign of the times for more orderly management but a recognition of the growth of trade and population. There were no railways, and roads were still bad, and here was a waterway leading into the heart of Suffolk which was an asset worthy of consideration.

The Commissioners proceeded to take stock of their powers and their first decision was to make a survey of the river. A Mr. Cawdren offered his services and to employ 'a machine of my owne invention for witch I have a Patent'. Banks needed cutting off to straighten the course, there were shaols to be moved, new [*Page* 158] bridges required and the channel to be deepened. A Mr. Langley Edwards was appointed surveyor and made an estimate that the work would cost about £3,000. The Commisioners agreed to raise the money by public subscription and by June 1759, the sum had been subscribed, with the names of Vanneck, Blois, Rous, Scrivener, Girling, Leman and Watling appearing on the lists. Tenders were submitted for various works and one by William Manning and James Collison was described as 'very extravagant and unreasonable'!

Langley Edwards appears to have been a consulting engineer and a very elusive man. Work had started on a bridge and the brickwork was not satisfactory. Edwards had left the job and was later found. He gave directions and then disappeared for a month and wrote that he had fallen off his horse 'and hurt himself much'. A breach occurred in the lock built near Wenhaston Water Mill and then the earth at the foot of each side of Blyford bridge was washed away and the bridge was in danger of falling. Where was Edwards? Why wasn't he available? The Commissioners were most indignant and summoned his assistant Samuel Jones to a meeting in February 1761, who reported that he had found his chief at Lynn and that he had had 'a fresh hurt that rendered him incapable of undertaking such a journey to the Blyth'. Later in the month, Edwards turned up and excused his absence from the previous meeting and asked the Commissioners to believe that 'his horse was taken lame on the road'!

The patience of the Commissioners must have been rewarded with this man's skill and advice! Samuel Jones his assistant, was still employed in 1765 and was then holding the office of Surveyor of the works of the river for which he received

£40 a year. In June of that year, the clerk reported that £4,000 had so far been spent on the navigation and the income derived was insufficient. Jones was selected as the first offering on the altar of economy and was asked to take a reduced salary. He refused and the [*Page* 159] post was given to William Bickers, a labourer from Wenhaston at £21 per year.

Bearing in mind the comparative value of money at that time – was £40 a year too much for the job or too much for the accounts to stand? This must have been a bad decision or the worth of the job was overstated.

Whether accounts were kept or what happened to them at this time, we don't know. On the fly leaf of the Treasurer's Account where entries started on 5th June, 1771, the clerk has recorded … 'there is no Treasurer's book or account among the River papers from the first meeting of the Commissioners in June 1757 for 14 years up to June 1771'.

Up to 1820, the records largely relate to the repair of the locks and bridges, the removal of shoals and making of new cuts. There are also the agreements on tolls to be charged for the carriage of merchandise, which listed, show the trade that once plied up and down the river:

Chaldrons of coal, culms of cinders and lime.
Lasts of corn, firkins of butter,
Weights of cheese, salt, timber, iron lead and groceries.

The average annual duties received from 1823–33 was £405 and the annual figures for merchandise for which duty was paid, was given as,

489 carts of chalk
9536 qtrs. of corn
4311 chaldron of coals.

The account of the Treasurer to the Southwold Harbour Commissioners for the year ending June 1836, shows some considerable increase in traffic.

Receipts. To dues on.

7078½ Chaldron of Coals	£707	17s.	0d.
29016 Qtrs. of Corn	£241	16s.	0d.
Sundry goods and wares	£142	19s.	8d.
1293 tons of ballast	£48	9s.	9d.
Tonnage dues	£3	9s.	6d.
The Nelson and Anne, for use of balks etc.		17s.	6d.
Interest allowed by the Treasurer on balance in hand		8s.	4d.
	£1145	17s.	9d.

[*Page* 160]

On the expenditure side of the account, I took note of one item – 'Salaries to Officers … £145.' Out of this sum, the Clerk, the Treasurer, the Surveyor and the Collector of Tolls were paid.

OH! happy days! Just think on the contentment arising out of accepted standards! But no! Our friend Samuel Jones refused to accept a lower standard and the Commissioners decided to accept a lower standard in the appointed labourer. You will recall the lines from the 'Deserted Village' –

'A man he was to all the country dear,
'And passing rich with forty pounds a year,'

This portrayal of the parson's wealth by Goldsmith, is contemporary with our period and no doubt the vicar of Blyford with his small parish by the River Blyth got less.

Just prior to 1820 the river commissioners were in frequent consultation with the Southwold Harbour Commissioners. When gales piled up shoals of shingle at the bar by the mouth of the harbour, ships could not approach and there was constant fear for navigation. At this time, one John Rennie was called in to make a survey and report on Southwold Harbour and the general state of the river. This was presented in 1820 and a third report on the state of affairs was made by 1841 and even to this year of grace in 1953, reports are still being made on this harbour.

In his report, Rennie decided that the main cause of the shallowness of the river channel was due to successive enclosures of saltings. [*Page* 161] He maintained that reduction in the volume of water flowing, had reduced the velocity of the water at its exit, thereby causing the river to deposit its silt in the riverbed and allowing the entry of tidal borne sand and shingle to remain at the mouth of the river. He informs us that the tide used to flow over 1100 acres now embanked: '550 acres of ancient embankments, 200 acres done 50 years ago (1770), 100 in 1780, 100 in 1804, 100 in 1807 and 43 acres in 1818'. 'Some of these are 5–6 miles from the mouth of the river, but most lie at Blythborough'. Rennie emphatically reported that if further enclosures were allowed, it would be necessary to provide a "receptacle … in lieu of the embanked lands, or the harbour will get worse". He also advised that the north pier should be extended 25 feet to give it the advantage of 5 feet over the south pier and that the channel should be dredged and made regular in width.

The Commissioners thought it a wise report, but its recommendations brought forth little action. In 1829, a Mr. William Cubitt was asked to make a survey and his report starts 'that it (the river) is now in a very bad state … and is capable of being much improved'.[15] The bar at the mouth still presented the same problem and Cubitt reports 'that there is a growth of shoals on both sides of the river … the small quantity of water passing over the bar failing to scour'. He also recommended straightening the channel and to removing projections to increase the flow. 'With northerly winds, the harbour is occasionally entirely blocked up for the whole width of the opening … and it remains closed until a change of wind takes place, when a cut has to be made by manual labour'.

There were more meetings – and the silt and shingle continued to pile up. In 1830 the original Act concerning the improvement of the harbour had to be considered. Originally passed in 1740 and the term continued by an Act of 1809 for a further 21 years, it became necessary to consider the extension of the legislation. A meeting to hear objections was called. It was said that [*Page* 162] the accounts were mismanaged and the harbour was in a bad state! A strong objection raised, was that masters of ships were by the Bill compelled to deliver an account of their cargoes. This matter however was already law and if omitted, presented an open gate for smuggling – a trade not unheard of along the Suffolk Coast of these times!

But for some years yet there seems to have been some laxity in the keeping of certain records on the river, but no doubt there was some explanation. In 1848, the Harbour and Railway department of the Admiralty wrote to R.B. Baas Esq., Clerk to the Blyth Navigation and asked for the reason of his failure to complete all of

15 W. Cubitt, a civil engineer from Ipswich, had completed two surveys and reports in 1820 and 1825 for a new cut in the River Waveney to accommodate the passage of ships from Norwich to the sea port of Lowestoft. See 'Map for the proposed navigation for ships from Norwich to the sea at Lowestoft' in William Cole, *A Poetical Sketch of the Norwich & Lowestoft Navigation Works* (Norwich, 1833).

the information required on certain forms. The Clerk's reply was not acceptable and from the admiralty he read, 'I am directed to state that there seems no reason why the forms are inapplicable for the insertion of the number of vessels annually navigating the river Blyth'. He was further told that for trade returns, he was required to record the quantity of grain and other commodities carried and 'a half yearly return made to the Admiralty in accordance with their Lordships' request'.

The Clerk noted on the correspondence – 'the records to be kept in a book'.

By August 1841, a third report on the harbour and the river had been presented by James Walker Esq., a civil engineer. He repeated the former arguments and reported that …, 'It is to be lamented that when the owners of the estates were perhaps balancing in their minds whether the land they could reclaim would pay the expense, they were not advised of the injury they were about to do to the public and themselves by the reduction of the backwater upon which the harbour is dependant'. But he agreed 'to take the encroachments as having had the sanction of time'. He described Blythborough bridge as 'a great nuisance' …, 'it is a disgrace to the Navigation and the County and ought not be to tolerated another year'.

[*Page* 163] We can get some idea of the state of the harbour from evidence given by one William Wayth master of the 86 ton 'Tyne' who appeared before the Commissioners at the Angel Inn, Halesworth on the 8th July, 1839. With the aid of 20 – 30 hands, it took him from the afternoon of Saturday to the following Thursday, to tie up at the harbour! He damaged and wore a new bawser which cost £15 and broke two big ropes and two others valued at £10: the vessel was damaged and he had to put hands to the pumps. Master Wayth reported that "the inside of the harbour is worse than the shoal outside".

In spite of the concern of the state of the navigation and the recommendations which were in reports, little action was taken and the reason appears as obvious then as it is to-day. In 1844 the meeting considered their costs which over a 10 year period up to 1844 averaged £256 per year. They thought the costs were reasonable but decided that all future extensive repairs should be done on contract. From 1848, when consideration was given to the amalgamation of interests of river and harbour, there is a disconsolate note of worry in the affairs of the Commissioners; the trade of the navigation was decreasing, but costs of upkeep continued. In 1856 the tolls taken were £90, 1861 £48, 1869 £79 and in 1871 £137 (probably as a result of a good corn year).

In 1862, John Cross of Halesworth wrote that the river bank of property recently purchased by him was in a bad state, and called upon the Commissioners to do something about it. The reply he received is somewhat significant of the state of affairs …, 'the Commissioners have no objection to repairs being done, but they will not do them'. By 1869, the Clerk and the Surveyor agreed to their third reduction of salary and in 1884 we find this worried body of men sitting in Town Room at Halesworth considering the advice of Frederick Meadows White Q.C., arising out of which, a notice appeared in the Suffolk papers.

> Notice is hereby given that this River (Blyth) having ceased to be used as a navigable river pursuant to the Act passed in the 13th year of the reign of Geo. II entitled "An Act for the making of the River Blythe …" the Commissioners have ceased to derive any income from the tolls or duties which have been accustomed to be received for the carriage of goods and merchandize and have now no means of keeping the bridges and the banks and sides of the river in repair, and henceforth decline all responsibility in respect of such repairs.

For the minutes of the 9th November, 1889 there is a solitary fact recorded. "The Collector reported that the sides of the river near Miss Crabtree's garden had given way". There is nothing further recorded; no full stop and signature of the Chariman! The consequence was no doubt disastrous for poor Miss Crabtree! Then there was an accident on the Old Chapel Bridge in Mells Hamlet. Two horses in a dray were going over the bridge, when one shied and the whole outfit went overboard into the stream. The men escaped uninjured but the horses valued at 120 guineas were drowned. The owners, the Southwold Brewery Co. threatened action, but the Commissioners merely quoted their public declaration on 1884 and dismissed any liability.

The bells had tolled – but the lying in state of this august body continued whilst its corpse continued to be defiled by actions. In 1894 the clerk wrote to the Board of Trade asking if they could afford any assistance to repeal the Act. He also wrote to the County Council asking them to take over the duties – and still the body could not be interred. In 1897 Herbert R. Stanford agent for the Blyford Hall estate complained that the bridge over the river at Blyford leading from the New Cut to the Water Boil bridge , which was erected by the Commissioners, was dangerous for cattle and carts to pass and requested that repairs be carried out in 28 days. The clerk was instructed to repudiate any liability!

[*Page* 165] The final minute of the old Commission is dated 14th June 1899 – 'resolved that the Commissioners dissent to the Mid-Suffolk Light Railway and that no reason be stated for doing so': and this is recorded especially for Mr. A.H. Liquorish, the present town clerk of Southwold who has by dint of circumstance only, inherited the 'body' of this railway and is still unable to give it decent burial!

Immediately after the first world war, farming still fared well and prices were maintained up to the early twenties. Then the urgency of home food production began to fade. By 1930 we felt the full blast of the economic depression which crossed the Atlantic and the state of farming affairs became rapidly worse.

Under the Land Drainage Act 1930 legislation was born for the land which was yet to pass through another decade of misery before the new law could really function for the benefit it was meant to give. Under this Act, the Catchment Boards were set up and in 1931 we see the East Suffolk Rivers catchment Board come into being which under schemes confirmed by Order, constituted Internal Drainage Boards. Although the River Orwell was a main river of the Catchment Board, land owners adjacent to the river preferred to remain free and continue with their own maintenance of the river walls. The River Stour (Essex and Suffolk) Catchment Board was separately set up in 1931.

Broadly speaking, the Catchment Boards were responsible (with a permissive right) for the repair and maintenance of the tidal river walls and the sea walls and other coastal defences, including the maintenance of sluices in these walls. (Internal Drainage Boards set up by the main boards were made responsible for the drainage of particular areas and took over the maintenance of the main drains in their areas.)

As an example, the River Blyth I.D.B. was set up on the 10th May, 1934, and held its first meeting on Saturday 20th June, 1934 under the Chairmanship [*Page* 166] of the late Earl of Stradbroke, with Messrs, F.W. C. Chartres, H.M. Cleminson, J.F.B. Ewen, E.T. Freeman, E.J. Le Grys, W.C. Mitchell, W.F. Neave, J.A. Peacock, S.W. Rix, M. Rouch and A.J. Groom present at this meeting. The first step for the board was to make arrangements for a survey of the area so that a rate book could be compiled. Each occupier within the board's area was assessed for rating on his land occupation. The first assessments were prepared by Mr. T.C. Clarke, surveyor to the Catchment Board and Mr. C.W. Flaxman was appointed rating officer. Mr. Willet Ram solicitor of Cross Ram & Co., Halesworth was appointed clerk to the board and held the office until his retirement in 1953 at the age of 79.

At the first meeting, the 'juniors' decided that the 'senior' body's attention 'be drawn to the danger in which the sea walls stood from need of proper attention and to the danger that they would be destroyed if there should be a specially high tide such as normally recurred every 30 or 50 years'. In July of the same year, they considered proposals from the Catchment Board for dredging the River Blyth in its upper reaches and the I.B.D. replied that in their opinion, the expenditure 'can be applied to better advantage in repairing walls of the main river, commencing near the outfall … rather than cleansing the bed of the river in its upper reaches'! At this time an estimate of £10,000 was given as the cost of putting the walls into repair.

This Board got down to its work thoroughly. Watercourses within its area were scheduled for attention and each member was allotted to take charge of certain district work and 'such money as is required be advanced to the persons in charge to enable them to pay for the work carried out'.

By April 1938 the Board came up against an old problem but one which was not to be tolerated if good drainage conditions were to be maintained, and so they minuted …, 'that all present obstructions to the free flow of water [*Page* 167] in internal watercourses within the drainage district erected by occupiers to prevent the straying of stock from marsh to marsh be removed by the board when such watercourses are next cleansed, and in future, any such methods would need the approval of the board'.

In 1944 consideration had to be given to the abatement of rates on land which had been flooded by salt water and which was temporarily put out of use. With the approval of the Catchment Board and the Ministry of Agriculture, the board invoked Section 24(7) of the Land Drainage Act 1930 and under certain conditions as set out, it was agreed to forego certain rates.

There are 21 Internal Drainage Boards whose Districts lie either wholly or partly within the county of East Suffolk and with the exception that their responsibility excludes the coastal and esturial defences, they are more or less counterparts of the old Commissioners of Sewers.

On the 1st April, 1952 the East Suffolk Rivers Catchment Board was dissolved and the East Suffolk and Norfolk River Board was set up in its place, under the River Board Act of 1948. Similarly, on 1st October, 1952 the River Stour (Essex and Suffolk) Catchment Board was superseded by the new Essex River Board – but the River Orwell affairs still remained in private hands.

THE SEA FLOODS 1953 IN SUFFOLK – DIARY OF OBSERVATIONS

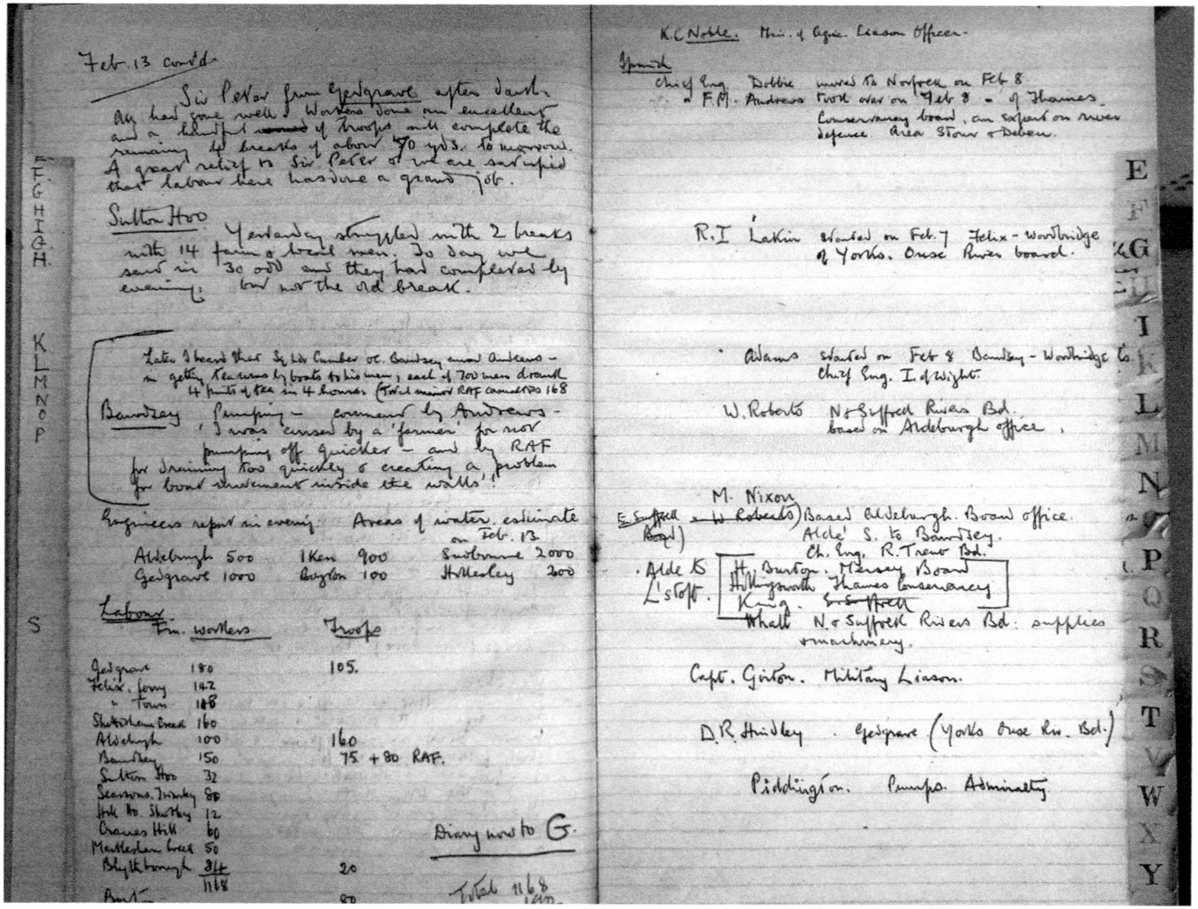

Plate 4. Sample pages from the diary of observations which Trist kept for a few weeks following the floods. Note how he organised his writing in this alphabetical notebook. For example, the diary filled section D (for diary, presumably) then continued in section G. Information about the engineers from the river boards, shown, was logically set down in between, in section E. (Suffolk Archives A2727/1/2) © Crown copyright

On the evening of Friday 30 January 1953 a fresh strong wind was blowing in the North but somewhat to the North East & rivermen saw to their moorings in the Deben.[1] By 8 a.m. Saturday 31st the wind had swung to North West and was strong. It blew all day and afternoon, intensity increased in strong spasms, by dark there was a roaring gale which kept up its intensity. I went to a hide for pigeons – it had blown down I put it up & it blew on top of me! Gale persisted in increasing force all night with terrific gusts reaching a crescendo like a jet fighter hurtling to earth in a dive. Walls shuddered and window frame shook. High tide in the Deben between 12 – 1 a.m. – I slept. On morning of March 1st, I looked out of bedroom window & the river had approached over Novacastria to within 200 yards of our house.[2] I could see the flood in the Pettistree Hall marsh. The roads were littered with branches. I went down to the river by the sandy lane to the side of the old harbour – debris was 6' above normal high tide mark. At 11.30 a.m., with still 2 hours before the p.m. high tide about 1.40 p.m., the water was only 2' below top of Wallers marsh wall with 6' in the marsh (the tide eventually stayed in the river all day). Flood on marsh & on fields of Cross Farm & Parkers – this was the small local picture – on the beach, houses flooded & timber & boats of Nunn's yard all in a muddle.

Later the paperman arrived with news from Martlesham – it struck there at the bridge in the darkness of early morning. Flood over the road between the Lion & the cottages beyond the P.O.[3] One began to imagine what might have happened up the coast – but I did not guess the magnitude of the disaster until more news came to me over the phone!

[*new page*]
(See new record on back of this page)[4]
February 1st/2nd – The gale – see notes in green book.[5]

Monday 2 February
Staff on survey & assessed 20,000 acres.

At Beccles NFU dinner at Lowestoft in the evening – the previous year whilst attending this dinner at this time I was called away to the fire at Fern Hill.[6]

1 This paragraph appears to be an introduction, written retrospectively. It is inserted before the first dated diary entry.

2 It was on 1 February, not March.

3 This is Martlesham Bridge over the River Fynn, equidistant from The Red Lion Inn, Main Street, on one side and the (old) Post Office, The Street, on the other, approximately 1 mile (1.6 kilometres) upstream from the Deben.

4 Written in the top corner, with a squiggled box around it.

5 Written in the space above the first ruled line of the note book, perhaps later.

6 At that NFU annual dinner in 1952, farmers lobbied the river board to 'unleash the potential of the marshes', voicing their frustration that 'all round the coast were marshes that used to feed hundreds of heads of stock … that were now derelict'. Criticising the old system of catchment boards, they argued that 'the marsh lands of the Suffolk tidal estuary were capable of enormous improvement, provided they had something more than a marginal board looking after them'. It was the farmers' hope that marshes would be brought back into production by the new river board if it 'would be able to take rapid and effective action' to keep water off the land. Anxieties about 'the terrible loss of growing land' due to the competing demands for its use were also expressed; the national campaign to increase food production clashed with local authorities' urgent need to build new homes in this area (*EADT*, 29 January 1952).

Tuesday 3 February
Did survey of Right bank Orwell with Lee.[7]

Wednesday 4 February
W. Jackson took Hayles & I up in his Auster from Ipswich airport.[8] Flew over Stour. Orwell. Deben. Ore & Alde. Up the Minsmere level & over the marshes south of Walberswick. Not good visibility & very bumpy 200–300' (see notes recorded in green notebook).

Thursday 5 February – [no entries]

Friday 6 February
Appeal in with press in Green 'Un for labour (see over page).[9]

See notes from flight by air – recorded on loose sheets in green loose-leaf and in 'cash' book.[10]

[*new page*]
See also notes in red field book now under [*section*] G.[11]

Friday 6 February
Desperate shortage of labour[12]
a.m. With Adams, Laurel Farm: desperate position for labour. Supplying his own men with neighbours help. Felixstowe Golf course under water – only direct access to the wall at Laurel marshes is up to the Ferry & on to the County Council wall – they are repairing their break. A DUK tried to cross the golf course but kept striking its rear wheels in the ditch – got over eventually conveying sand bags and log timber.

3 breaks in wall opposite golf course & some beyond the Council cross wall.

The gravel heaps over the road – wooden huts piled up & distress of belongings lying about – (see notes under [*section*] G).
Further diary of February 6th from Felixstowe Ferry.[13]
On night of 1st February, Holm Hill was surrounded by flood: Mr & Mrs King were marooned in their cottage at Kings Fleet sluice. Mr & Mrs Brundish, foreman Holms Hill marooned & were taken off by A. Adams' men by boat.

Alf & Chas soon mounted on Suffolks on the Laurel marsh rode out across the flood & tried to drive 2 horses standing over their bellies in water – they were too frightened and would not drive, so got a boat and chased up behind them.

At Ferry Boat when we halted at 1 p.m. for a drink to warm us out of a bitter north wind, Alf hesitated & then told me he had no breakfast. Inside we saw the 4' 6" tide mark cut in the doorway of the bar – wonderfully cleaned up & the landlord in good cheer. The Adams in great spirit against desperate odds and shortage of

7 Harry Lea, the district officer assisting Trist.
8 Trist later writes that the pilot was 'Mr R. Jackson'.
9 Written in large square brackets over four lines.
10 This flight is described in 'The great sea floods of 1953', above, pp. 7–10.
11 Added at the top of the page with a large arrow, cross and tick mark.
12 Written small in the margin.
13 This text was written in Section F; for continuity it is included here.

labour, battled against gaps under their own supervision after engineer Larkin had generally directed for the centres up to top of wall.[14]

[*text in section D continues*]

Saturday 7 February
Office all morning organising & phone afternoon at Waldringfield.
Meet Engineers in the evening.
Launches weekend appeal in Green 'Un.
Replies through with ½ hour & continued to 11.30 p.m. Over to Reg re admin.

Sunday 8 February
& overleaf. (Sunday) 1.50 a.m phone from Police – a worried wife whose husband on Bawdsey pump not yet home.

Phone started before 7 a.m. – with volunteers –- they met buses in Ipswich, came in cars & bikes.

At Petistree Hall 9.30 a.m. & drove down to the wall. Warner in charge for the estate, with farm workers & 80 volunteers from Ipswich & surrounds doing a job with their spades, unusual to them and in appalling muddy cold conditions. A useful job which locals completed on Monday night. At nightfall from Bawdsey, a long trail of buses, vans, lorries & bikes full of tired mud bespattered men.

The real attack on the gaps had just begun.[15]

Evening to Woodbridge to join Hayles around the quay in a search for motor boats.

[*new page*]

Sunday 8 February
Bawdsey:[16]
RAF – troops – filling bags laying electric light on poles over the breaks. Small boats ferrying wooden stakes inside the wall. Pontoon craned up Waldringfield quay, lorried & launched on the marsh at Bawdsey - Hayles as O.C. Boats.

8 fire engine pumps with hoses over the walls. 3 Ministry pumps.

(Hayles & I pontoon) re. boats – tide lowering all of the week in the river up to Thursday, 12 February making landing of men & materials up to the work difficult.

By late afternoon, the cold work intense & then a bitter wind & first snow – the constant tramp of hundreds of feet along the walls, especially over the wet mud of the break areas – became mucky & pumps on the wall, fire hoses,

Petistree Hall:
80 Town volunteers with boots, spade & food by bus car & cycle gave Sunday work – people not used to river winter cold & mud. Organised from an appeal in Evening Star by C.A.O. arranged also with Bristow on telephone to a late hour.

(refer to previous '49 flooding of these marshes & area plugged)

---

14 R. Lakin of the Yorkshire Ouse River Board.
15 There are squiggles next to this text.
16 The underlining is as Trist used it.

At evening – Bawdsey – on the wall as light fell, an endless trail of men stretching out into the dim distance in a wriggly line of the wall, tired men mud plastered, with shovels, plodded their way back after a days digging in mud. Marshes drained of water.

Lowestoft & Yarmouth:
3 breaks in Breydon Water filled over weekend of 7–8 February & also 1 break in river Waveney north of Rackhams, The Rookery, Carlton Colville. At Lowestoft tide came over the road by car park & Wallers Cafe into Oulton Broad & up Oulton Dyke & flooded surrounding marshes of Somerleyton & Oulton.

Pumps at work & they now wait to pump & sluice off.

[*new page*]
Appeal to farmers for volunteer labour in E.A.D.T.[17]

Monday 9 February
South end of Gedgrave peninsular 570 yards wall damaged: 100 yards down to salting level, rest 1' 6" above salts: [*By the*] 9th only 200 yards done. From Lower Gull (Flagbury Point?) running south along from Dock Farm to Boyton Hall, 11 breaks & tops.[18]

Gedgrave break – position away from Gedgrave Hall at end of long mud track & then 400 yards over a causeway over the mud to the wall.

Visit of Minister of Agriculture & Home Secretary to Felixstowe - met by Sir R. Gooch, Ridley & C.A.O. at Martlesham, accompanied by Lord Cranbrook & Colonel John Hare.[19]

Sunset at Gedgrave, the marshes from Orford to Fagbury Point at mouth of the Butley Creek were still full of water & the little woods stood out as islands, & tops of tall thorn hedges.[20] Beyond lay Orford Ness with Kings & Lantern as full of water as could be held – a searchlight beam moved.[21]

The breaks in Boyton wall – 11 as above – were centre filled today, leaving front & back battering.[22] 145 farm workers for 10th February, as from the press appeal.

Hayles on Deben – boats 'Master of the Boats' (River Deben).

Boyton to day 70 farm workers, forestry men & troops.

Executive Committee met at 11 a.m., I remained to give a report of conditions and action taken and retired to my phone.[23]

---

17 Written at the top of the page with a large, squiggly underscore.

18 Trist rightly queries his reference to a Flagbury point. His confusion is perhaps a sign that tiredness was beginning to take its toll. Fagbury point is on the Orwell west of Felixstowe, near Trimley marshes, and Flybury point is on the Lower Gull near Orford Ness. In this instance, Trist is writing about Flybury point.

19 4th Earl of Cranbrook, John David Gathorne-Hardy (1900–78), was honoury air commodore of the No. 3169 Suffolk Fighter Control Unit, Royal Auxilliary Air Force, between 1950 and 1961 (also an archaeologist and a founder and the president of the Suffolk Records Society, in 1958). John Hugh Hare, 1st Viscount Blakenham (1911–82) was a veteran of the Suffolk Yeomanry, MP for Sudbury and Woodbridge, 1950–63, and vice-chairman of the Conservative party, later becoming Minister of Agriculture, Fisheries and Food.

20 Flybury point.

21 King's marsh and Lantern marsh, Orford Ness.

22 Battering in this context refers to the shaping of the land between two different levels. Batters are the side slopes that connect the wall to the contour of the surrounding land.

23 This text is marked by a large, open square bracket.

At 12 [*noon*], Sir R. Gooch & Ridley & I to aerodrome at Martlesham, met Sir Thomas Dugdale & Sir David Maxwell Fyfe, Home Secretary, who had been to see things at Felixstowe – & flew off to Hendon and on to the house.[24]

Heavy rain overnight. Apart from cold & fine snow of Sunday afternoon – this is the only set back so far.

Pettistree Hall wall finished by local labour after Sunday aid of town volunteers.[25]

[*new page*]
155 workers out – the start of a force which swelled to 1400 by Friday this week.[26]

Tuesday 10 February
Lakin says Laurel Farm to just north of Kings Fleet are 50% up to 2' of wall. 80% of breaks in Kirton–Falkenham area are in hand. Assault of saltings by Kirton break with landing hands off the salts – Hayles.

Bawdsey–Shottisham:
Andrews estimates 130 million gallons water – weeks of pumping. Walls all up to 2' of top from Bawdsey to Ramsholt, but top to finish & no attention yet given to front batter and back & back scours: 1700 troops on this wall & how the mud bulged from the bags as they tramped – the Engineer Andrews coming off the wall late, sank over his knees in bag mud whilst crossing a filled break.[27]

156 farm labourers in response to CAO appeal & sent as under:

Gedgrave 55, Boyton 34, Shottisham 17, Bawdsey 40, Sutton Hoo 10 (2 new breaks).

Orford: & Boyton Hall area – Orford Ness – Waltons story of the heifer on the lav! [28]

60 yards to south of Town break filled on the old horse shoe bend. Sluice which was damaged is temporarily sealed. Too few men on the Sudbourne wall, but access very difficult –Engineers struggle to get pumps in operation at Poplar Farm, Iken. Water a little down, some drowned beans showing at Cowell – 2 pumps at work.

Gedgrave:
Our farm labour has started to turn the scales of anxiety. Sir Peter had the highest praise for them – came from all over Suffolk, farmers paying their own men and in most cases providing & paying for the transport of buses & cars.[29]

Operation – Boots, Spades & Grub – the message to everyone over the phone.[30]

Aldeburgh:
The big break south of Town marshes is temporarily abandoned for contractors. Sent a plan of the scour, it is 7' in centre, cannot be done for 3 weeks. Labour now on heightening an old wall on the south west of the Town

24 Hendon Aerodrome and the House of Commons.
25 There are lines in the margin next to these two sentences.
26 Written at the top of the page with a large, squiggly underscore.
27 This later addition was written further down the page with its intended placement indicated by an arrow.
28 The text after 'Orford' is a later addition in different ink, in small writing. This story of the heffer evidently amused Trist because he included it in his typescript.
29 Sir Peter Greenwell, on whom see above, p. 20, note 72.
30 Next to this line are marks in the margin, for emphasis.

[*new page*]
to prevent more tide coming into back of the Town via the Town marshes.

Ministry of Transport are anxious over the growing fresh water flood at Latimer Dam.

Benacre Pumping station cannot work for days. 200 yards of the 100 river leading to pumping station is full of sand & gravel – an excavator tracked yesterday along the dunes for the Kessingland Holiday Camp.[31]

Wonderful response to the EADT appeal to-day for Wednesday men. NFU Saturday Ipswich.

Did tour with Ripoll: Aldeburgh office – Iken – Sudbourne a sea – the flattened sand of the arable around Iken Church – the pale colour mark of the tide – the drowned kale near Crag Farm: tide debris piled up on the arable indicating the highest mark. The hedges were decorated with pieces of reed & straw & weed.

[*18 blank lines*]
Rain by late afternoon & during night.

[*new page*]
Hayles O.C. Boats. Deben.
a. 3 pontoons
b. 3 motor boats
c. 2 whalers
d. 2 barges

[*new page*]
See also copy of Noble's report of 11th[32]

Wednesday 11 February
A.E.C. labour organised from all over Suffolk:

Gedgrave	140
Felixstowe Ferry	94
Felixstowe Town Hall	103
Shottisham	14
Aldeburgh	70
Sutton Hoo	7
Ramsholt Lodge	43
	471

A strong wind in NW heavy overcast in morning with rain near. Rain after 11 a.m. in strong interval showers – snow & sleet by nightfall.

31 River Hundred.

32 Written in a different ink in the space above the first ruled line, at top of the page. The numbers of men and machinery deployed were routinely reported to the Ministry of Agriculture (MAF 157/25, Reports and surveys on extent of land flooded in East Suffolk).

Report from Andrews 10.30 am on Ramsholt Bawdsey line, all bar 3 are safe. These 3 suffered heavily last night by numbers of troops crushing out the mud in the bag. Break no. 13 at Shottisham Creek was 79 yards long & only got at by 300 yard walk along the wall or by boat.

To day the Deben is 1'3" higher than predicted.[33]

Kirton Creek (with Andrews):
North side. 4 breaks 26,10, 23 & 25 yards all up near top: 14 yard top off near sluice & a 40 yard on south side of creek on the top of the concrete: tops repaired. In this creek 400 Navy plus local farm workers have been on & in 4 days have done a huge job – bags laid out very well with a base of big concrete lumps all hand carried out along over 100 yards of wall. Water almost all off the arable behind – The sodden land still with huge puddles & ditches full I think looked worse than the flooded land!

Levington Creek:
Andrews said it could not possibly be done until after the spring tides. The break in west side is 29 yards. Across the front is a width of 6' over which some new wall can be erected, but behind the scour from centre of the wall goes from 6' to 10' at rear of the wall. If an attempt is made now, no work can be done until dead low tide as the water is still draining off the 40 acres behind – low tide is 9 p.m. – in the morning, there are still more urgent areas with greater stretch of land behind which must have attention.

[*new page*]
Trimley marshes:
The assessment is 1000 yards of top of wall gone.

[*6 blank lines*]
Adams and his gangs have sealed just beyond the Kings Fleet. Boats are operating north of this point. Adams labour supplemented to nearly a 100 & he was pleased!
River Orwell, Right bank:
5 at HMS Ganges are sealed. No work going on – tops on Hill House Farm, Shotley (Wrinch) 5 p.m. between Hare's Creek–Collimer Point.

Break below Oyster beds Cranes Hill large break. ¾ filled by 11 a.m. 60 boys of HMS Ganges working. Water still on 40 acres.

Marshes opposite Over Hall, Shotley, small gangs working on breaks.

[*6 blank lines*]
River Stour:
South of Erwarton – at Ness Hall break excavator at work on 3 breaks.

Frank Keeble was quietly getting on with his own work.

Engineers meeting at the Bull Woodbridge: all emphasis to night on safeguarding Felixstowe Town by the cross wall being thrown up behind Trimley marshes & safeguarding sewers.: the priority of R.A.F. Bawdsey

See above continuation.[34]

33 There is a cross in the margin next to this sentence.
34 There is a cross mark in the margin.

also priority of Aldeburgh. Two days ago I saw the plan of the breach in the centre of the Town marsh wall which is 7' deep & will be left for 3 weeks for contractors: the Town labour force & our supplements are concentrating on the wall at back of the Town. A comment at meeting 'do these first & give every attention' … 'the rest is merely agricultural land'.[35]

[*new page*]
Did not attend Engineers meeting at Bull this evening[36] –

Thursday 12 February
Weather
More rain overnight. Very cold intermittent rain & sleet throughout morning. Soon after 2.30 p.m. a fine snow which increased in intensity as darkness drew on and intense cold.

a.m. office: Phone calls pouring in offering labour from all over Suffolk, fear of oversubscription. Decide to take no Sunday labour from any source.

Private walls.[37]

Dixon of Bidwells re. Wolverstone Estate trying to help on his own, promised to double his labour from 50 to 100 on 13th to go to Shotley Church for repairs to breach below oyster beds at Cranes Hill & tops. 10 for Hill House farm for tops & 100 for Searsons Farm, Trimley.[38]

Martlesham Creek:
South side breaks & top off sluice wall. 30 yards, 4 & 11 yards breaks and 20 labour on these to day. [*illeg.*] and no work is started. Boards men are filling 3 breaks at the top end of the Creek. We arranged 50 men for Friday. Ripoll went there after lunch & found a handful of men struggling on their own and a River Board foreman in charge – he said he could do with 200 – we got him 50. Labour reported to my house & office from 6.30 a.m. and by 2.30 p.m. we had over 1100 men mustered and transport ready with buses to report for tomorrow.

Blyth river:
The colossal work south of Aldeburgh had occupied all – we were getting to a surplus supply of labour & I thought of the poor Blyth which for so long had suffered 486 acres of its marshes under the tide. In touch with Norwich they said get some to White Hart Blythborough – by night we had 83 for them.

---

35 This text was written in a ruled off space, earlier on the page. Here, it has been added in the place indicated by a corresponding cross mark.

36 Written small in a different ink at the top of the page.

37 Written as a note in the margin. The issue of whether poorly maintained river walls in private ownership contributed to the flooding in some areas was discussed at the inquest on 36 of the 41 Felixstowe flood victims. It was pointed out that owners of marshes were under no obligation to maintain walls as a public service, only in so far as was necessary to protect their own land (*EADT*, 18 February 1953).

38 David R. Dixon of Bidwells gave evidence at the Felixstowe inquest. As a surveyor with the firm responsible for maintaining the stretch of wall on the River Orwell protecting the marshes (belonging to Trinity College, Cambridge), Dixon described various breaches in the wall, one of which was 40 yards (36.5 metres) wide. It was his opinion that the wall was adequately maintained, bearing in mind the standard of maintenance in the district, and the purpose of the wall. At the time of the flood, he said the owners were negotiating with the river board to take over responsibility for maintenance (*EADT*, 18 February 1953).

Bawdsey:
700 R.A.F to day:
9.30 p.m. a fairly strong 'high' wind in the tops of the trees & intermittent blasts of strong gale. A searchlight thrust a bright beam into the sky from Bawdsey and in our farmyard covered with snow, the light became a blur from this light which accompanied me & my lantern as I went over to Pickles stable with bread.[39]

In the evening the BBC broadcast hourly tide reports from the M.O.A.[40]

B.B.C. Tide 1'8" above prediction.[41]

[*new page*]
Reports from evening of 12 February.

700 troops standing by for emergency call out if walls break over weekend. Bawdsey R.A.F. men to stand by in their own area. Nixon to be up at dawn on Sunday by helicopter base on Sudbourne – to fly the whole coastline & reporting by radio to ground – having troop dispositions known & 15 call up positions where troops are required.

Engineers are to patrol their own banks as from daylight on Sunday morning – but warning given to them to beware of being cut off on the bank should there be breaks.

Weekend concentration Aldeburgh, Felixstowe, Bawdsey, Gedgrave.
Farm workers wherabouts

	10th	11th	12th	
Gedgrave Hall	55	140	230	
Boyton Hall	34			
Felixstowe Ferry		94	160	for Laurel Farm
Felixstowe Town Hall		103	100	for Town marsh
Shottisham Creek	17	14	360	
Aldeburgh		70	90	
Bawdsey	40		142	
Sutton Hoo	10	30	45	
Pettistree Hall		43		
Searsons Hall			50	
	166	494	1177	
			494	
			166	
			1837	= 3 days

39 Presumably 'Pickles' was a pony stabled at Trist's home in Waldringfield, about 6 miles (9.5 kilometres) upstream from Bawdsey.

40 Ministry of Agriculture.

41 That night, people along the flood-stricken East Coast reportedly spent a long vigil by their radio sets listening to BBC flood warnings, after the first high-water reports had heralded the beginning of the cycle of spring tides that would last until 19 February. Headlines warned about 'Danger Tides' and newspapers printed tables of predicted high tide times at Lowestoft, Southwold, and Felixstowe for the next few days (*EADT,* 12 February 1953).

[*11 blank lines*]

12 February
Made report to M.O.A – & PD. PLC.[42]

[*new page*]

Friday 13 February
Potter offered tractors & trailers to Edward Nuthall the Ipswich contractors who are to take on Shottisham–Ramsholt.[43]

Martlesham Creek:
3 breaks completed to-day. Marshes at Shottisham Creek breaks completed to stage 1. Total of 200 yards, nearly all with back deep scours & main damage to rear of wall. The line is now a series of horseshoe bends, excavator at work has been filling in behind the bags. Tuesday & Wednesday work with small local gang under Warner (foreman of farm on estate.) Thursday & Friday supplemented up to 160 men. These marshes clear of flood to-day.

Ramsholt Lodge Marshes:
North of pub from North–South – 2 tops 4 yards and 9, 9, 14, 13, 11, 9, then 44 yards with 2 small 'rouths' of a few yards of bank.[44] Breaks in yards to foot of wall with scours behind – 65 yards.[45] Here & at Shottisham the water lashed the inside of the marshes & did much damage. The burst threw huge lumps of clay across the marsh. Ditches full of water, but main flood off to day.

Bawdsey:
See note overleaf also[46]

The great assembly of cars, vans, fire engines & troops has almost vanished. 2 pumps have been at work and half of the gate posts in the marsh now show – about 2'6" of water lower than last Saturday the 8th. Main force today on 150 yards of break on Poplar Farm marshes opposite Falkenham Creek – the back of stage 1. for Bawdsey–Ramsholt is about complete. Leaving Ramsholt Lodge (as above) to do tomorrow.

Weather
Fine snow off & on most of morning. Snow lies on the arable & all is a winter scene. Wind died away p.m. & it seemed warmer – but perhaps it was the work. Simmons

42 The PD was the provincial director of NAAS at Anstey Hall, Trumpington, Cambridge.
43 Possibly Potters of Framlingham and Woodbridge, the Fordson tractor dealers. Their prominent advertisements in the local newspaper promised farmers, 'the utmost priority to repairs to Fordson Tractors and Implements which have been damaged in the Flood Areas' (*EADT*). The contractor, Edmund Nuttall of 22 Grosvenor Gardens, London SWI, was contracted for works on the left bank of the Deben at Shottisham Creek (MAF 157/25, Reports and surveys on extent of land flooded in East Suffolk, Ipswich flood report no. 6, 14 February 1953).
44 The meaning of rouths is not certain. Some claim 'routh' is a Yorkshire word for a section of rough land.
45 Written small in a ruled-off section.
46 Written in the margin and boxed.

& I parked at Pettistree Hall & walked to end of the Ramsholt Lodge wall.[47] The river was calm & on the turn. At Ramsholt Lodge the stock bellowed as usual and flocks of pigeons made the customary scene.

[*new page*]

Friday 13 February continues
Sir Peter from Gedgrave after dark: All had gone well. Workers done an excellent and a handful of troops will complete the remaining 4 breaks of about 70 yards tomorrow. A great relief to Sir Peter & we are satisfied that labour here has done a grand job.

Sutton Hoo:
Yesterday struggled with 2 breaks with 14 farm & local men. To day we sent in 30 odd and they had completed by evening: but not the old break.

Later I heard that Squadron Leader Cumber O.C. Bawdsey cursed Andrews – in getting tea urns by boat to his men, each of 700 men drank 4 pints of tea in 4 hours. (Total minor R.A.F. casualties 168.

Bawdsey:
Pumping – comment by Andrews – 'I was cussed by a "farmer" for not pumping off quicker – and by R.A.F. for draining too quickly & creating a problem for boat movement inside the walls'! [48]

Engineers report in evening: Areas of water estimate on 13 February.
Aldeburgh 500, Iken 900, Sudbourne 2000, Gedgrave 1000, Boyton 100, Hollesley 200

Labour

	Farm workers	Troops	
Gedgrave	180	105	
Felix Ferry	142		
Felix Town	118		
Shottisham Creek	160		
Aldeburgh	100	160	
Bawdsey	150	75	+ 80 R.A.F.
Sutton Hoo	32		
Searsons Trimley	80		
Hill House Shotley	12		

47 J.E. Simmons, assistant CAO (Executive) supporting Trist. Evidence that the CAO staff were working all hours is in the flood report which Simmons prepared this day. After typing a two-page summary of activity for the Flood Emergency Division, Whitehall, he signed off, 'P.S. Sorry this reads incoherently – but its night starvation!' (MAF 220/9, Correspondence between East Suffolk AEC and MAF Floods Emergency Division, Whitehall).

48 Marked by a large open bracket, probably highlighting the story about the boat movements that is repeated in 'The great sea floods', above, p. 21.

Cranes Hill	60		
Martlesham Creek	50		
Blythborough	84	20	
	1168		
Boyton		80	
Poplar Iken		170	
		610	+ 80 R.A.F.

Diary now to [*section*] G[49]
Total 1168 [+] 690 [=] 1858[50]

[*section 'G' – continuation of the diary entries*]
See diary [*in section*] L for Noble stats. of Equipment[51]

Saturday 14, February
p.m. inspections

Martlesham creek:
50 acres. 3 breaks.[52] South side. Hill Farm marshes 3 breaks topped to wall & good work, men working on mending the top of low wall by the sluice facing the Deben. ⅓ marsh still under water: lying in the flood, neatly stacked piles of concrete slabs, a spade still stuck in the ground and an iron wheel barrow, where the Board had been at work on maintenance. Opposite the marshes of Sluice & Sink lay as a former reminder.

Kyson Point: Woodbridge wall.
4 breaks & 5 tops or back scours. The people of Woodbridge got to work on Monday & carted loads of earth. The iron seats & the shelter on the wall had stood the test. Marshes now empty of water but ditches full. As I walked the bank it was low tide and on its edge teal and widgeon hovered & whistled, whilst shelduck cackled – gulls soared & dived – a clumpy flapping as a cormorant left the water & took off on an upstream course. As I left the wall the whine of powerful wings & 3 swans with outstretched necks and in line ahead flew towards the Tide Mill.

Sutton Hoo:
At the entrance the cottager trying with a small engine to pump out the little flow by his house which on the Sunday has got into his downstairs rooms. Up to the big house a good view of the river, but no one at home![53] I had not yet met Mrs Barton, but her voice in desperate appeal was heard the previous day on the phone by several of us, 'our little bit marsh is in a forgotten corner of the river, we have

49 Written in bottom right-hand corner of the page.
50 Sum in the corner of the page.
51 Written in different ink in the top right-hand corner of the page, underlined.
52 In margin.
53 Sutton Hoo House had been the home of Mrs Edith Pretty, near the site of the archaeological excavations for the Anglo-Saxon ship burials in 1938–39. It was later renamed Tranmer House and acquired by National Trust.

instructed our own men & foresters, ... please send some farmworkers![54] We sent 32 & as the sun was low in the sky on Saturday with snow cloud, I walked over the sodden marsh to the walls. 3 breaks & 7 tops & scours completed up to top height and lovely work.

Opposite lay Woodbridge seemingly quiet from river view & behind shoppers were busy buying their Saturday foods, whilst genial P.C. [*gap in text*] formerly of Stowmarket directed the traffic by the Crown.[55] Rooftops covered with snow. As I got to top of the wall, [*new page*] the sentinel curlew of a 'flock' cried out & up they got. The teal 'sprang' – the redshank piped warning, whilst the shelduck just turned to look and waddled off at ease on the mud. Some of Ponds coke was in the top of the wall, part of 200T of coal & coke washed out of his yard opposite.

As the sun was sinking in a great red ball over Kyson, I turned to face the hill of Sutton Hoo & the derelict hulk on the saltings reminded me of the Buried Ships grave on the hill – of past history.

Leaving by the hill along the river I looked down at the 2 old breaks of 1942 & the flooded x acres – now bare & muddy the scene of former disaster and a forerunner of the sight that is to come along the coast. I retired to tea 5.30 p.m.

After a fortnight, this was another D-day. Tides were high again – but all went well. On the Deben, my guide was the high tide at Ramsholt 'Dock' on the opposite bank and 2 miles below Waldringfield. To night the high tide is 1.00 a.m. one of the most convenient times of the 24 hours of any day! The emergency organisation staff are on call on the phone night & day. Teleprinter Ipswich – Ministry – radio communication – helicopter dawn patrol.

[*11 blank lines*]
after Saturday 14 – quote NAAS staff employed with the Board[56]

Saturday 14 February continues
a.m. in office. A meeting to decide duty phone rota. Action of labour for next week. Red X. WVS. YMCA. [*illeg.*] Transport. Sorting of accounts. Contact with contractors. Lord Stradbroke phoned of pleasure expressed of all the way, by HRH Edinburgh who visited Bawdsey, Friday. Late morning, calm & river tide OK.

[*new page*]

Sunday 15 February
Weather. Calm. Snow still lying, slight fall before noon. Tide 12.30 p.m. Ramsholt 13' odd. River full but calm. No phone over night & staff had leisurely taken over as we changed from phone duty.

1.10 p.m. County Magazine BBC. At Sun Inn, Woodbridge, Mr C. Westripp who recorded 'he had come up to Woodbridge for a quiet job' & didn't bargain for all this in his pub.[57] Programme introduced by Ralph Wightman. Fred Martin of Moat

54 Lieutenant Commander J.L. Barton and his wife acquired the estate at auction in 1950, not long before the floods.

55 The Crown Hotel on the corner of Quay Street and Thoroughfare, at the crossroads with Cumberland Street and Church Street.

56 Written in different ink in a box in centre of blank area, marked by a cross.

57 Though reticent, it appears that Mr. Westrip embraced this opportunity. At Woodbridge Brewster Sessions on 12 February 1953, as licensee of the Sun Inn, he asked for an extension of his hours to

Farm Clopton in majestic deep if full Suffolk voice & got stringer down to sing beer & barley: Taken up by Mr G. Adams of Walton, a maltster 'in the old days we had beer before breakfast & at 1 ½ a pint!' Fred continues 'he's just making a living' – (just 'making out' as he usually says when I meet him at the Bull!) Ralph questions Fred Martin on rabbits who says 'I like to leave my rabbits to my farm hands' & Archie J. Wellings of Beccles A.E.C. pests says 'that isn't good enough' & is off onto method. 'I put my money on the good ferret & a good dog' says Ralph & Welling agrees and boasts of 400 rabbits in 3 weeks with ferrets.

Landlord of the Dooley, Felixstowe, 'I used to be up at 5 baiting the horses & into bed at 9 at night'.[58] When I first heard of the pub vacancy 'I said I've enough friends about there and this beats muck spreading, so I took it'. 'In those days, apart from chaps off the grain barges, we had few visitors'. Mrs Morris, wife of farm labourer, one of a family of 14, 'life is what you make it' – a philosophy of sound sense just at this moment of disaster.

Mr Cutting of Aldeburgh has worked in his village store for 50 years, it was established 1872 'we can do anything now from a box of matches to a sweet of furniture' – 'used to sell 000s of lamp glasses at 2d. a time & we still buy pegs from the gypsies'.[59]

A pleasant interlude in almost the 1st opportunity to sit back and relax, whilst at the time the tide was full in the Deben, but calm as I saw it on leaving the Maybush at 1 p.m. to go home with a clear conscience of a calm river running.[60]

[*new page*]

Sunday 15 February continues

Benacre Sluice:

Opened at 2 p.m. Shingle cleared from the 100 River & excavators also at work on the shingle at the mouth of the old sluice. Work going ahead on the pumps & may be ready in 2 weeks.

11pm for the coming Saturday, to accommodate a BBC party in his bar. Reportedly, the chairman of the hearing asked, 'Is this the real BBC?' He was told that the BBC were indeed recording their Country Magazine show in the pub, to which he replied, 'Oh, I thought they might have been down here because of this high tide. Any how, you can have your extension.' (*EADT*, 13 February 1953).

58 The Dooley Inn or Walton Dooley, was also known as the Ferryboat Inn in Ferry Lane, Walton.

59 The store, A. & M. Cutting, was in Front Street, Mendlesham, near Stowmarket (not Aldeburgh). A gypsy peg bought from this store in the 1930s is in the Museum of East Anglian Life: 'A gypsy would call at the store to ask Mr Cutting whether he needed any clothes pegs, and he would often order a gross (12 dozen or 144). The gypsy would then make these over the next two or three days and return with them, often being paid in goods rather than cash' (https://abbotshall.wordpress.com).

60 This broadcast featured Robert Irwin, 'singing another of the old Suffolk songs that BBC producer Francis Dillon continues to revive'. Other participants were, 'Mrs K. Norris, of Dallinghoo, wife of a farm labourer, Mr A.J. Welling, a pest officer and warrener, of Beccles, Mr C.F.W. Miles, landlord of the Ferryboat Inn at Lower Walton, Mr C.E. Cutting, who keeps the village store of Mendlesham, Mr H.B. Furlong, a well known expert on old clocks, from Oulton Broad, Mr G. Adams, a foreman maltster from Walton and Mr Fred Martin, of Moat Hall Farm, Clopton' (*EADT*, 14 February 1953). Understandably, there are small variances between Trist's account of the names and places he heard in the broadcast and those reported. However, the Dooley and the Ferryboat Inn are one and the same.

[*16 blank lines*]

Monday 16 February
A clear sky and a warm sun of an early spring day. Snow still lay, but melting. No wind and the tide rose calmly. The helicopter patrol reported no trouble: it landed on the beach at Ramsholt at 2.30 p.m. At 12.30 p.m. to Ramsholt Arms & met Hayles. Walked up stream marsh by Keepers Cottage, Ramsholt. 5 breaks & 7 scours men still at work. This marsh drained. More E.V.W.s from Elvedon.[61] Ex A.E.C. workers on Ramsholt Lodge wall. Took photos of effect of sea beet roots lifting concrete flags on this wall.[62] Flights of widgeon over the river. Still 2' water on opposite bank in Paul's Corporation marshes, where last Sunday week February 1st, it was level with the eaves of the barn on the marsh. Men working south of Falkenham Creek seen in the distance. At a break below Ramsholt Lodge a terrier's backside in a hole & out he came with a rat in his mouth, shook him & banged him on the ground – rat bolted & jumped into the delf & swam under water, up for a breather & down again. Terrier spotted him swimming for the land on opposite side of the hole by the delf & went to wait his landing & caught him. Retiring to Ramsholt. Arms, looked up at Ramsholt Church pink in the bright light surrounded by its few houses, a kingfisher flashed by!

North to River Blythe:
Water had travelled over the Blythboro' road nearly up to Wenhaston, The Reydon & Tinker [*marshes*] were more than ¾ full, some grass highlands [*new page*] showing below Reydon Hall. Called on G. Girling and discussed the events. At Southwold the scene along the front of the harbour was one of devastation. Shingle had filled the road about 6' deep. Bungalows, houses & huts were shattered. Boats and heavy sea buoys were thrown over the beach & road into the marsh behind, 4 fire pumps were at work by the bridge over the Blyth opposite Walberswick & 2 fire pumps at bottom of the Town pumping out the marshes.

17th & 18th, tides behaved. No flood incidents of breaks, but tragic loss at Aldeburgh. (See EADT of 18th).

Final meeting of Engineers at the Crown Hotel, Woodbridge 8.30 p.m. 18th February.[63]

---

61 Under the European Volunteer Worker (EVW) scheme, the British government sent officials from the Ministry of Labour to camps for people displaced by the Second World War to recruit workers in order to meet the need for labour in key occupations. One scheme, called 'Westward Ho!', was designed to attract people to work in agriculture. In all, more than 80,000 men and women came to Britain as EVWs. They received the same wages as British workers, although the British government considered the EVWs to be economic migrants (hence 'volunteer workers'). It is likely that the EVWs that Trist writes about were working at Lord Iveagh's Elvedon Estate, a very large mixed farm pioneering the use of mechanical harvesting in the early 1950s.

62 Sea beet, sometimes called 'wild spinach', grows wild on shingle beaches, cliffs, and bare ground near to the sea, as well as in saltmarshes. It can be cooked and eaten.

63 After the threat of further flooding receded, a letter was published in the local newspaper, from the chairman of East Suffolk AEC (R.C. Ridley) and Trist: 'Sir – On behalf of the East Suffolk Agricultural Committee we would like to express our grateful thanks for the assistance so readily given by all who helped in the repair of the sea defences in the recent disaster. Whilst the situation was being assessed, several hundred servicemen, together with a handful of other workers, were positioned. The first civilian volunteers came from Ipswich, and a group of 80 men did an excellent job throughout a cold Sunday. Subsequent appeals made through the columns of the "East Anglian Daily Times" were immediately taken up by the N.F.U. Offers of help poured in immediately from all

Tuesday 17 February
Staff meeting re. flood survey, Melton.

Wednesday 18 February
N.F.U. D & E Committee, Ipswich[64]

Thursday 19 February
Aldeburgh:
Absolute havoc on the beach at the south end of the town. A mass a bulldozers & excavators & mountains of shingle. The break at Slaughden had pushed the beach flat on to the marsh – now bulldozed back. Marshes full of water & tide still coming in – wall of Town marsh badly damaged. All breeches filled in centre except the one with deep scour (& scene of Tuesday tragedy & loss of 1 life) Boats cast up on the saltings and lieing on their side. Houses near the Slaughden end still had sandbagged doorways.

Hazelwood marshes still flooded on the tide from an unsealed break.

Minsmere
East Bridge. Some water still under Scotts Wood & much in the Nature Reserve. Water came over the wood on to the Theberton marshes.

From Dunwich Cliff
The old clay wall always covered with shingle in which the Tamarisks grew, had been recently under extensive repair by the Board – the sea came over the beach and over the top of the wall and pushed thousands of tons of [*new page*] gravel on to the marsh. Bulldozers at work pushing up a wall of shingle on a new line with the wall which goes south after the join of the Coney cross wall. The latter wall, 102 yards, shattered to the base, on which 5 men toiled with spades & forks. A break in the wall south of Coney x [*cross*] wall about 100 yards down, of 25 yards right to base of the wall: 4 men working there.

Poor Minsmere's fete!

[*16 blank lines and a six-day gap in diary entries*]

Wednesday 25 February
Gedgrave Hall:
Water off but marshes very bogged & ditches full of salt water.
[*next entry is five days later*]

Monday 2 March
Nuthall's at work, excavators & bulldozers a total of 11 breaks – 4 on east side about 40 yards, 2 near Orford quay, the rest along the south wall totalling 595 yards. & 2 scours of 45 yards – almost 700 yards total.

---

over the two Suffolks. The response of the farmers was magnificent and typical of the East Anglian's reply to a call to arms. The farm workers who toiled in the worst of weather and in deep mud can be proud to know that it was their efforts which saved many acres from further flooding from this week's tides.' (*EADT*, 23 February 1953).

64 NFU D & E Com. was the NFU Disaster and Emergency Committee (Ipswich Branch).

On March 2nd I soil sampled at Richmond Farm with Sir Peter. The arable on the marsh still lay absolutely sodden. Wheat was absolutely dead & the beans quite black. Marsh grass showing signs of going black – further time will show death! The delf ditch was not harmed, only silted up where clay was blown out of the wall breaks. Culverts had been opened up and some water furrows cut to let off water – the ditches are still full of salt water.

[*9 blank lines*]

[*new page*]
Southwold
The sea came over the shingle beach south of Easton cliffs and followed a line north of Buss creek over to the Reydon marshes. Also came over in the weak places in the shingle foreshore north of the harbour. Bulk of damage to the walls was done on the southern leg of Buss creek, where it faces Reydon marshes and where the tide was driven into the wall from the north west.

Southwold–Easton Bavents – saw gravel over 10 acres, feet deep covering the marsh. Bavents cliffs 20–30 feet high, badly eroded & large recent falls. (See paper from Boggis). 1 ½ miles – say 10–20 acres gone in 50 years). Large conglomerate pebble etc. stones. torn out from foot of cliff & sea bed. (A area, vital for sea protection to the road entrance to Southwold and the whole of the Blythe river valley to the North.)

Benacre Sluice.
Shingle cleared from river and sluice opened at 2 p.m. on February 15th.[65]

Gap torn in dunes above the sluice extending to the coastguard station about 80 yards wide. Sand & shingle washed back into the marsh 200 yards and the Hundred river completely filled with sand and shingle 300 yards up stream. The causeway between the water filled gravel pits leading from the Beach Farm to the sluice was badly damaged and traffic could not get down.[66] In the vicinity of the sluice there was such devastation of sand covering, that an area of about five acres had completely lost its old character.

(Also break in the shingle pit at the south end)[67] Pumping station flooded & much damage to engines – at work again by the end of March, with the fall end of the river now deepened & almost twice its previous width. This was a problem area as the fresh water piled up in Kessingland level and flowed back onto the main road at Latimer Dam several times after the sea water had gone off.

To the south as far as Benacre Broad the rest of the Denes had been flattened back and had been bulldozed to a height of 10' for [*blank space*] yards.

[*new page, written on the first page of section* 'H' *in the directory notebook*]
Walberswick
The sea had previously been over the marshes – in fact several times in the past ten years. The great shingle bank looks much the same, although perhaps a little flatter on the top. Immediately out from the track to the beach, by the high ground around

65 Written in above the heading and the main text.
66 Cross mark in margin.
67 Cross mark in margin.

which the Dunwich river flows, the tide had exposed the actual sandy earth of the mainland by working off the shingle. It is likely that Southwold took the main brunt at this part of the coast and the jutting harbour helped Walberswick front (although much flooding in the houses near the shore) there was a large breech in the shingle bank further south about opposite Little Dingle Hill and not far from the breach of 1949 which flooded the Point & Westwood marshes as the break in the Point wall had not been sealed. This was done by the Catchment Board in 1951 and through many months of 1952 the Board were dredging the Dunwich river from Walberswick right through the Dingle marshes and were within ½ a mile of Dunwich when this tide caught the excavator on the marsh!

Havergate Island: 3–4000 dead rabbits – one alive on the wall. – from a local in The Crown, Orford.
Orford Ness: Mr Walton – the heifer on the seat!
Humberstone Farm: cattle brought up the road, b'ditch on either side – by [*illeg.*] walking up to his chest in water – one lost.
The sandbagged railway arch.

[*section* 'K'– *indexed as* 'Flood borne debris']
Record of flood borne debris
High House & Middle Barn Bawdsey: At Middle Barn a solid piece of marsh 70 yards long, over 25 yards wide and 2½ to 3' deep was deposited at the back of the marsh about 50 yards from the gateway opposite the entrance into Bawdsey manor grounds. Other huge pieces lie in the ditch & at top of the marsh below the lane leading down from High House to the marsh and weigh 000s to 000s of tons. It is likely that they are the edge of the ditch running through the Queens fleet: the clay earth has the complete matt of grass, reed & sedge roots. From the position in which some of it is lying, ie. flat on the marsh, it was floated by the tidal wave from the river and crossed over a 4' 6" post & rail fence doing no damage – it is therefore likely that the depth of the water in the marsh was 8' at least. More lies over the crumpled iron railings on the side of the Ferry road.

Searsons Farm Trimley: 24 March, 1953. Met B. Smith and Dixon of Bidwells
The debris here has also been torn out of the edge [*of*] the ditch through a fleet. It has no clay and is made up entirely of rotten fibrous matter with a course grass top, some with reed and all with a dense matt of sequoia roots – estimated total about 5000 tons.[68] It is 2 ½–3' deep – one block is 200 x 55 yards. The tide height in the thorn hedges on the edge of the marsh is 10'.

A few hares were seen and a mole had made a new run on a higher sandy part of a field of beans; the top ¼ acre of which was not flooded, where perhaps the mole took refuge. This arable is shooting thistles yellow coloured: docks in a strange pinky red, mayweed and creeping bent is again growing. The general expanse of the lower marshes, look horrid in a dirty brown death and the white of salt is starting to appear with the characteristic red orange colour of earth in the low places. Green plovers and mallard called out, be gone, this is our land to be undisturbed for several years.

68 Sequoia is a giant redwood conifer first introduced into Britain in 1853 and actively planted by the Forestry Commission during the twentieth century.

Also on Kirton Lodge & Corporation Farm Kirton – debris out of the Corporation Fleet.

[*section* 'C' – *indexed as* 'Crops*'.*]
Winter Beans
10th day under water – outer edge of leaves seen turning black. As soon as water is off, the whole flower is black and dead. A susceptible crop.

White turnips
Tide washed for a few hours – leaves on the uppermost tide line showed an opaque scorch and leaves lower in the tide were withered. A very susceptible crop, succumbing quickly.

Winter Wheat
Distinctly hardy under flooding up to 14 days plus. Die back of tips and half the length of blade from 10–14 days flooding. Dead at 3 weeks.
[*section* 'F' – *Originally indexed as* 'Fodders available*' and amended to* 'Restoration of drainage*'*]

Fodders available
~~Farmer~~
~~The Gables Tunstall 1 stack baled straw at Tunstall~~
~~1 stack clover~~
~~W.Suffolk AEC 100T. Baled wheat & rye straw at Lakenheath~~[69]
Organisation handed over to the NFU.

Restoration of drainage
Clearance of clay washed out of wall breaks and other flood borne debris in delf & farm ditches. Estimate at 13th April '53 of value of work to be done by farmers or hand labour, on ditches & restoration of bridges
517 chains, £2,372 [70]
Work to be done by contract excavators £6392 – 1215 chains
This length includes 300 chains of Internal Drainage Board streams.
Excavators at work on 13th April – 9 [*in number*]

[*section* 'E' – *indexed as* 'Engineers*'*]
K.C. Noble Ministry of Agriculture Liason Officer.

Ipswich

Chief Engineer Dobbie	moved to Norfolk 8 February.
F.M. Andrews	took over on 8 February – of Thames Conservancy Board, an expert on river defence. Area Stour & Deben.
R.I. Lakin	started on 7 February – of Yorkshire Ouse River Board. Felixstowe–Woodbridge.

69 The text is crossed through but still legible.
70 Chain, an imperial unit of measurement being 22 yards or 66 feet or 4 rods (20 metres). Ten chains make one furlong and 8 furlongs, one mile. Therefore 517 chains is about 6½ miles (10.4 kilometres).

Adams	started on 8 February – Chief Engineer Isle of Wight. Bawdsey–Woodbridge
W. Roberts	started on 8 February – Norfolk & Suffolk Rivers Board based at Aldeburgh office.
M. Nixon	based Aldeburgh Board office. Aldeburgh south–Bawdsey. Chief Engineer River Trent Board
Aldeburgh to Lowestoft	[*in square brackets*]
	H. Burton, Mersey Board
	Hollingsworth, Thames Conservancy
	King, ~~East Suffolk~~
	Whall, Norfolk & Suffolk Rivers Board: supplies & machinery
Capt. Girton	Military Liason
D.R. Hindley	Gedgrave (Yorkshire Ouse River Board)
Piddington	Pumps. Admiralty.

[*section* 'L' – *indexed as* 'Areas flooded & gen. stats.']

Sea Floods 1st Feb 1953
Final Records made by County AEC staff

	Acres [*flooded*]		
River direction land to sea	Arable	Grass	Total
R. Stour Left bank	185	259	444
R. Orwell Left bank	224	750	974
R. Orwell Right bank	132	307	439
R. Deben Left bank	277	1380	1657
R. Deben Right bank	727	1196	1923
*Bawdsey–Shingle Street	112	623	735
R. Ore /Alde Right bank	2122	2795	4917
R. Alde Left bank	70	758	828
Aldeburgh–Walberswick	29	1585	1614
R. Blythe R. & L. banks	146	1815	1961
Southwold / Lowestoft	8	1049	1057
R. Waveney Right bank	2	3958	3960
	4034	16475	20509

* 62 orchards included

Duration of flooding

			County Totals				
			Arable acres		Grass acres		
0–12	hrs	(828)	123	(1514)	7	(2342)	12%
12–24	"		274		258		
1–2	days		170		368		
2–4	"		261		881		
4–7	"	(3206)	327	(14961)	1340	(18167)	88%
7–10	"		522		1435		
10–14	"		1013		2150		
14–21	"		1267		3880		
Over 21	"		77		6156		
			4034		16475	20509	

[*Pre-flood utilisation of flooded acres in East Suffolk and other counties*]

	East Suffolk	All counties in G.B.
Tillage under crop acres	1448	23966
Tillage bare	2100	23085
P. Grass	16475	99053
T. Grass	486	12746
Orchards	62	827
	20,571	159,677

Livestock Losses

	County Total	Deben
Cattle	146	120
Sheep	303	303
Pigs	86	18
Poultry	1289	1114
Horses	11	7

Flooded Crops

	Samford	Deben	Blyth	L'land	TOTAL
Wheat	58	981	4	27	1045
Oats	17	72			89
Beans	100				100

Temp. crops	12	473	1		486
Kale		21		3	24
Horti. crops	56	3			59
Other	36	73	16		130
					1933

[*other agricultural losses*]
39 partial damage to buildings
5 total damage to buildings
42 tons fertilizer flooded
34 tons bulky fodder flooded
56 tons concentrates
40 mangold clamps
10 other clamps

Flood report to Ministry of Agriculture by Noble of 14 February, 1953[71]

Civilians	Military	Dozers	Pumps	Draglines	Craft	DUKW
6	120	3	1	1	4	
20		1	1	1	2	
20		1	2	1		
		1	2	1		
100	100	10	1	1		
100	70		1	7		
50	80		2	1		
160	160		1		3	2
20						
150	400					
100		1	3		8	
100	400					
100						
100	75					
140				1		
170				2		
115	360				3	
100		1		1		
30						

71 A record of the number and type of personnel, machinery and craft in use for wall repairs, submitted to Whitehall.

30						
72						
150						
30						
83					–	
1946	1765	18	14	17	20	2

1765 = 3711

Acres still flooded 7000 approximately.

There are more notes in this document that have not been transcribed, e.g. telephone lists, and information about damage to sea defences that is more complete in other documents.

SEA FLOODS 1953 – REPORT

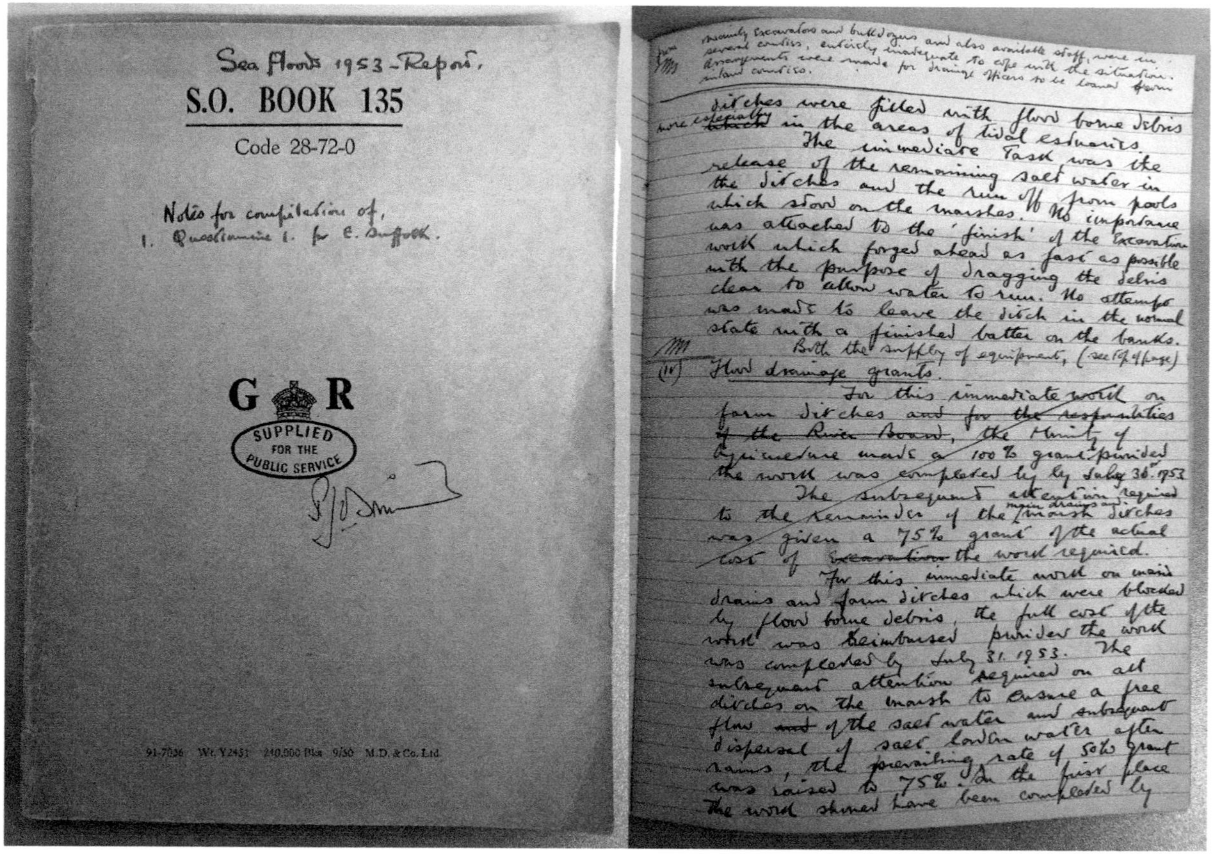

Plate 5. The cover and a page from the notebook in which Trist drafted sections of a MAFF report on the effect of the floods on agriculture. Note the addition of section headings, later insertions of text, and pencil annotations in this working document. (Suffolk Archives A2727/1/3) © Crown copyright

Question 1. – East Suffolk

1) The survey of the flooded areas:

On 2 February, a team of district advisory and drainage officers were allotted sections of the coast and river estuaries and asked to make an immediate rough assessment of the area of flood. This information was to hand by 4 p.m. Owing to difficulty impossibility of approach to the river walls, no account of any value could be made on the extent of wall damage. On 3 February, the County agricultural office chartered a private aircraft and flew up and down each estuary and the coastline from Brantham on the River Stour to Walberswick and plotted the position of wall damage which was submitted to the County flood organisation H.Q.

Following this survey, district advisors were instructed to collect full details of the area flooded, crops and livestock lost etc. for each farm concerned in their district. The flood line was plotted on district 6" maps and subsequently plotted on scale maps together with plottings of all wall breaches, scouring and over topping of sand and shingle defences.[1]

Following completion of the land survey, advisory officers proceeded to take pilot soil samples and gradually intensified sampling down to single marshes in the case of flooded arable and approximately 50 acre groups for grass marshes.

2) Assistance by A.E.C. to River Board:

a. Loan of staff
b. Loan of equipment
c. Organisation of labour
d. Organisation of river craft
e. Contact

a. Loan of staff:

Two drainage officers were attached to local River Board offices and four were assisting in the field in charge of pumps. Two advisory officers were attached to the southern H.Q. office at Ipswich together with two clerks and one clerk was loaned to a stores depot.

b. Loan of equipment:

8 lorries were put at the Board's disposal for the transport of troops and other labour organised by A.E.C. A large quantity of small hand tools were supplied, together with 8 excavators, a low loader and complete personnel for operation.

c. Organisation of labour:

35 excavator drivers and other drainage field workers were immediately loaned. Accomodation and catering facilities were made available at the Debach E.V.W. camp which was reopened to receive 750 troops.[2] About 100 men were mustered from local labour exchanges. Arising out of a press appeal, 80 town volunteers from Ipswich were organised for wall work one Sunday. Following a further press appeal to all farmers in the two Suffolks, a corps of skilled farm workers increased each day until February 14. Some transport was organised by the A.E.C. in the form of

1 See Maps A2727/2/1–16.

2 RAF Debach, a few miles north of Woodbridge, was used from late 1945 to 1948 to accommodate German prisoners of war and later displaced persons working in agriculture (known as EVWs), under the control of the CAEC. The camp was abandoned in about 1948.

buses, but the majority was provided and organised by farmers loaning their own men, who provided their own spades and rations. The volunteers started at 156 and rose to over 1300 on the twelfth day after the flood and something over 3000 men days were provided to the Board in one week. The greatest part of the labour and transport costs were borne by farmers.

For allocation of this labour, the A.E.C. obtained a phone report of the general disposition of all labour which included troops, which throughout the day was supplemented with detail of extra help required in certain areas. In this way, A.E.C. staff were able to direct offers of labour to specific points on the coast. At the end of the day, the Board's office at Ipswich received a full report of the numbers of men to be expected at each point on the following day. In a number of instances, the difficulty of contact with the Board's telephone and the lateness of the hour, made it imperative for the A.E.C. to direct the labour on their initiative and even to meet and give supervision to the workers.

Arrangements were made with the N.F.U. to see that farm workers received special flood relief extra rations. In addition the W.V.S. were asked to provide mobile canteens. [3]

– Army & R.A.F. personnel
– from labour exchange!

Consider a note on the

– excellent work of the farm workers
– the willing answer for help from farmers
– [*ditto*] Forestry Commission
– W.V.S. on canteens
– surveyors department of county council.

[*8 blank lines*]

d. Organisation of river craft:

In the early days of the flood, the repair to wall breaches was made difficult by the problem of sea flood water preventing access over the marshes and a broken line of wall. The worst area affected was in the river Deben estuary where the total length of the width of wall breaches was over half a mile. For several days, access was by gradual approach from one breach to the next after a temporary sand bag repair had been effected. A large quantity of sundry small equipment, timber and hundreds of thousands of bags full of wet clay were carried by hand along the walls to the breaches, until the high tides had receeded to allow mud to be dug out of the saltings near the breaches.

To assist the operation of transport on the river and inside the flooded marsh a fleet of little ships was organised by the A.E.C. On first enquiries for the loan or hire of motor boats it was found that practically all of them, inspite of being laid up or under repair, had had their electrical equipment flooded.

Hired 3 motor boats, 2 whalers, 3 pontoons and 2 barges.

3 Women's Voluntary Service, now the Royal Voluntary Service. A record of the significant contribution this organisation made in a multitude of ways can be read in a booklet produced by the WVS concerning flood relief work, 'Report on Help Given by East Suffolk W.V.S. after the Storm and Tempest on January 31st–February 1st 1953' (Suffolk Archives, HD 1848/2).

3) Meetings of Engineers - River Board & A.E.C. at Crown Hotel.
A.E.C. to engineers
– routes to approach marsh
– high saltings for craft approach
– arranging car pilots for tours
– boats for transport
– disposition of labour

4) Liason for dead animals – Local Authorities:
See Felixstowe surveyors report October 9th, '53 & Digby's report.[4]
Both on Civil Defence file.

[*10 blank lines – there is no ' 5)'*]

6) See report under 7.

7) Flooded crops stats – see attached sheet

8 & 9) – see stats on file. 36/79

[*legible notes related to the questionnaire end here. The next facing page appears to begin with a draft contents page for the MAFF, NAAS report, 'The Effect on Agriculture of the East Coast Floods, 1953', written by the working party set up by the main advisory committee of sea flooded land, on which Trist served.*[5]]

[*new page*]
<u>1953 Flood report</u>
[*tick marks made against the list in a different ink, suggest the content was referred to and checked off at a later date.*]

<u>Part I</u>
a. [*ticked*] The causes and effects of storm and tide
   letter to Dorey of 3rd May[6]
   (Extracts from Civil Engineers Report if permitted)[7]
b. [*ticked*] General descriptive account of the flooding
   (edited from CAOs reports)
c. Questionnaire 1. – compilation of county reports.

4 J.H. Digby, a livestock husbandry adviser, wrote to Trist about a slaughterman collecting dead animals off the street in Felixstowe (MAF 157/25, Reports and surveys on extent of land flooded in East Suffolk, Memo from Digby to Trist, 3 February 1953).

5 According to notes made by Trist, the report was not published by the ministry until 1962 (A 2727/1/8, P.J.O. Trist papers, folder of notes and pamphlets on salinity and reclamation of soils). However, a draft was in circulation by 1959 to which Hilda Grieve makes reference in *The Great Tide: The Story of the 1953 Flood Disaster in Essex* (Chelmsford, 1959).

6 O.G. Dorey, chairman of the MAF working party set up by the Main Advisory Committee on Sea Flooded Land to examine the effects of the 1953 East Coast Floods.

7 Institution of Civil Engineers, 'Conference on the North Sea Floods of 31 January/1 February, 1953: a Collection of Papers Presented at the Institution in December 1953' (London, 1954).

Part II

Formation of Floods Emergency Division

1. [*ticked*] Considerations given to the technical aspects – (5)
Drafted.[8]
2. The supply of gypsum
3. [*ticked*] Diversion of excavators and loan of DWS staff [9]
   see letter to Dorey of 3 May[10]
4. Coastal Flooding (Emergency Provisions) Act 1953
   Annual acreage payments
   Method of assessment and procedure
   Rehabilitation measures for losses & repairs
5. Drainage work and grants
   Lord Mayors Flood & Tempest Distress Fund[11]
6. (Simmons to record)

[*11 blank lines, then a footnote*]
Referred to Dorey on 2 May 1955 – to Johnson for
Pt. I. Permission to reprint.
Pt. II. Will Land Drainage do the record.[12]

[*new page*]
Gypsum
24th February meeting[13]
March 1953. Discussions with Sc-Watson, Prof Zuur & Prof Deloffre.[14] Zuur visited Essex and was of opinion soils were much heavier than in Holland and that Dutch results of experimental work would not give us good guidance. Deloffre consulted on gypsum experiment.

Prior to above – a committee met on 27 February – agreed that there was simple means of determining quantity of gypsum required for all types of soil conditions. Agreed for tillage where analysis contained -1% or more NaCl in 6' – 2T good gypsum per acre – owing to restricted supply, or 2½ if the by product containing 20–25% moisture from ICI fertilizer plant at Billingham: allowance to permanent and temporary grass was not agreed necessary, except in cases of complete destruction – allotments to be supplied via Lea.[15]

8 In the margin.

9 DWS: the Drainage and Water Supply division within MAF.

10 A notation added at the edge of the page.

11 Lord Mayor's National Flood and Tempest Distress Relief Fund, established on 10 February 1953 and financed by public subscription (matched pound for pound by the Treasury). The fund was the primary instrument through which people, including farmers, received financial assistance for flood-related losses.

12 G.C. Johnson, provincial director, MAF (MAF 220/9, Correspondence between East Suffolk AEC and MAF Floods Emergency Division, Whitehall).

13 In a different ink at the top of the page.

14 Under the chairmanship of Professor Sir James Scott-Watson, chief scientific adviser to the Ministry of Agriculture, officials with knowledge of the problems involved with land restoration were invited to discuss their recommendations in order to formulate advice for farmers. Participants included experts from Holland and France (professors A.J.Zuur (1902–1961) and G.Deloffre) as well as British agricultural scientists and advisers. This informed a pamphlet circulated by the NFU to its members (A2727/1/8, MAF, *Advisory Leaflet on Treatment of Land Flooded by Sea Water*, March 1953).

15 Imperial Chemical Industries was a British company which, for much of its history, was the largest manufacturer in the country. Defunct since 2008.

Committee also recommended experiment and Dr Rye & Davis were asked to submit proposals. (Refer to see Chemist's report).

Dr Schofield of Rothamstead advises that whilst he felt 2T per acre would be insufficient in some cases – bore in mind the supply position.[16]

Discussions on cultivating in gypsum. General agreement in light harrowing to ensure an even dressing and to counteract any bad drainage effect with uneven land. (In practice little harrowing was done, owing to time and wet land.)

Delivery free at farm in all cases where approved by Committee after consultation with soil chemist.

DAO [*computed?*] tonnage required.[17]

Report to Ministry with information of nearest rail /[*illeg.*].

Free transport was arranged by [*illeg.*] from station to farm, whilst farmers made own spreading arrangements.

Quantity issued in 1953–4–5.

[*two lines on the page indicate a section break.*]

Considerations given to administrative and technical aspects
ON 53/18
MAF 3209
1st meeting 24 February – 27 February
3215 – warnings & leaflet on treatment of S … 3229
In March – Sharrocks paper – Guidance in general from soil chemists – FL/AS/14/ Pizer – County advice on cropping and cultivation – much advice [*illeg.*] from Holland.
3236 Committee on Coastal Flooding
– terms of reference and committee members
3rd meeting – discussions and sorting out of information
– proposals for experiments & trials
4th [*meeting*] 23 April – gypsum controversy
gypsum experiment & long term experimental proposals
4 June – Pizer proposals.
5th [*meeting*] of 10 June – Planning of experiment sites.
Proposals of grass land, legume, experiment
for observation plots and gypsum. [*fruit?*] trials. average crops for long term.

Quote list of Published Papers eg. FL/AS/23 Pizer & Davis & Dutch extracts.[18]

FL/AS/27 Sat Com [*illeg.*] analysis method
FL/AS/28
Losses / & Lord Mayors Fund Transfer of responsibility December 1953.

16 Rothamsted, the British agricultural research institute also known as Rothamsted Experimental Station, the Institute of Arable Crops Research and Rothamsted Research.

17 DAO: District Agricultural Officer, the first point of contact for farmers, with counties being divided into several smaller districts.

18 This is written as a footnote in a ruled-off space.

[*new page*]
IV[19]
Coastal Flooding (Emergency Provisions) Act 1953
MAF 3240 26 March
MAF 3244 a) Acreage payments. See CAEC 53/47 – The 'moves' and describe in ordinary language.[20] and 53/40
CAEC 53/29 b) Other rehabilitation measures
1954 payments. See CAEC 54/21
1955 [*ditto*]

IV.C.I.[21]
Section V.C.I. – "Acreage payments to farmers & allotments"
Act – see above.
The act provided for assistance to farmers small holders and River Boards to rehabilitate flooded land and to enable the speedy reconstruction of coastal defences and see paragraphs. 3 + 6 of MK 4.
The act gave power (see MK 1. MAF 3240 MK 2.) 1953 Payment details.
Mk 5. 1954 [*ditto*] 1955 [*ditto*]

[*8 blank lines*]
V.C.2 – "Other rehabilitation methods"
MK. 3
MK. 6

[*On the next double-page spread is an incomplete table for* 'County losses'. *These are not transcribed here because the 1962 MAFF report includes the complete figures.*[22]]

[*new page*]
Section III
Emergency measures adopted
The local action taken by County Agricultural Committee staffs immediately the disaster had occurred reflects the value of the county organisations in special circumstances. Whilst the maintenance and repair of coastal defences are, for the most part, the responsibility of the River Boards, the size of this disaster was something which no ordinary organization can normally take into account and in consequence it was found that technical and working staff and equipment available to the Boards could not possibly cope with the situation. It will also be recorded elsewhere that apart from help given by county agricultural committees, many other organisations assisted in a multitude of ways.

a. The survey of the disaster

The immediate task of agricultural committee staffs was to supply a quick assessment of the areas flooded and the damage done to sea defences to Ministry headquarters. At the Land Drainage division of the Ministry, the chief Land Drainage Officer [*gap left here*] Johnson C.B.E. was responsible for the assessment of River

19 In red pencil, in top left-hand corner of the page.
20 Insertion of text indicated by an arrow.
21 Underlined in red pencil.
22 A summary is in the Appendix, below p. 115.

Boards' needs and through engineers appointed to take charge of operations in flood areas, the necessary requisitioning and delivery of equipment was arranged.[23]

Both the staff engaged and the detail collected in the survey of the disaster, varied from county to county according to circumstances where the flood areas were large, most of the technical officers employed together with drainage and field officers of the A.E.C. and workers of the Land Service.[24] In some counties it was possible to complete an immediate survey of details, whilst in others, the flood covered such a wide area that the task of immediately recording all facts was a physical impossibility.

The first action taken by all counties was to map the fullest extent of the flood line on the land and its furthest approach up ditches. In Kent, where 9,300 acres of water still lay after 3 weeks, a survey of the flood line was made weekly and mapped.* [*footnote in text*] [25] In Lincolnshire (Lindsey) the county staff already had the experience of the 1947 fresh water flooding which followed the thaw of a 6 [*week*] period of snow and with the assistance of the National Farmers Union compiled an extensive record form, which was sent out to farmers, and which was subsequently completed and checked in the presence of A.E.C. staff and committee members. The information being subsequently used as a check against the claims made by farmers on the Lord Mayors Flood and Tempest Fund. In this instance the county staff and the N.F.U. had previously worked together and repeated the operation. In other counties, the N.F.U. were of immense assistance in other ways, whilst in other counties where the disaster was small, the committee managed without their help. These facts are recorded to show one of many instances where local decisions to cope with local problems are frequently best level to those on the spot.

The question of recording the number of actual breaches in the sea and river walls together with general damage to the top and rear walls, was a problem on account of access. In some marsh areas, the distance of flood water was often one to two miles from dry land to the sea wall. The only approach to survey damage to a wall was by walking the wall as far as the first breach. Whilst this was the particular concern of the River Board for repair, the county A.E.C. was trying to assess the need for labour requirements.

In East Suffolk an aerial survey * [*footnote in text*] was made from an Auster aircraft two days after the storm and damage to river and sea walls was plotted and subsequently reported to the River Board.[26] This information proved to be extremely helpful in many ways. The majority of engineers in charge of flood zones had been loaned by drainage boards all [*over*] the country and in most cases found themselves in a strange countryside with no knowledge of the best access to the

23 Eric A.G. Johnson, chief engineer, Land Drainage.

24 Agricultural Land Service part of the Land Division of MAF.

25 [*Trist's footnote reads*] * (Note to go at foot of page on the above * occurs) Flood water also lay in Norfolk, Suffolk and Essex over very large areas from 3–7 weeks, but no weekly mapping of the water line was carried out. Soil analysis has since shown, more especially on heavy soils, where the flood duration was lengthy. In addition, in the summer of 1955 the continued growth of salt tolerant flora if studied by species and quantity in relation to other flora was a good guide to prevailing salt content.

26 [*Trist's footnote reads*] * It is recommended that should other floods occur where the size and scope cannot be easily assessed from the ground, that a helicopter is made available for the senior officers of the River Board and the A.E.C. to make a joint survey of the situation. This immediate liason enables these two authorities to appreciate their respective responsibilities in the disaster and to enable immediate consideration to be given for the necessity of both labour and equipment.

marshes. There was not time for them to travel and find what they considered was the best route and with the record of breaches and local knowledge of the marshes, the East Suffolk staff was able to be of considerable assistance to the engineers in piloting them to the breaches, and considering the shortest accessible route for the conveyance of materials. In addition, the very large labour force organised by the A.E.C. could be directed to areas for repair as was arranged with the engineers.

After immediate preliminary survey of the extent of the flood, samples of flood water were taken at various places on marsh levels and at the head of river estuaries, to assess the range of salinity in the water. After a few days, when consideration had carefully been given to all the facts which needed recording both from a technical angle and the subsequent consideration which would have to be given to losses of livestock, stores and equipment, county staffs were allotted sectors of the coast and river estuaries to record the following details:

1. Name of farm flooded with details of ownership and address of farmer if different.
2. The acreage of arable and grassland flooded with the acreage of each crop and bare land flooded.
3. Soil type.
4. Past cropping history, depth of ploughing and observations on the immediate reactions of crops to short duration flooding.
5. The duration of the flood water (records being subsequently added).
6. Condition of drainage.
7. Flood debris.
8. Losses of livestock, stores, equipment and machinery: loss or damage to stacks, clamps and farm buildings.
9. Loss and damage to gates, fences and roads.

[*new page*]
Continued from overleaf

Aerial photography
The R.A.F. undertook a complete aerial survey of the flooded coastline, but several counties recorded that very little use could be made of the photographs. Some were taken at too great a height to be of useful interpretation to the layman unversed in this study and in general the photographs recorded large areas showing the extent of the flood rather than indicating detail which would have been more useful. The weather at the time was very dull for good photography and possibly high winds were a deterrent to low flying.

[*12 blank lines*]

[*new page*]
b. Assistance to and liason with River Boards: staff and equipment.
[*the first paragraph is crossed out*]
The normal staff and equipment carried by River Boards was so completely unable to cope with the situation without assistance, that it was immediately obvious to the A.E.C. that their help was required in any shape or form. This varied according to flood circumstance and the pool of staff and equipment available.

Where committees were still running a machinery contract service, crawler tractors, winches and trailers were loaned, whilst the bulk of useful equipment in the form of ditching excavators, bulldozers, lorries and low loaders were loaned from drainage departments. In addition, spades, shovels, forks and a variety of small tools were transferred in great quantity.

In most counties it was found that excavators could only be loaned for a brief period of a few weeks owing to the importance of returning them to the land to clear ditches filled with debris. In some areas, none could be loaned to the Boards on account of the immediate problem of clearing sand and other debris from the mouths of main dykes which were holding up the discharge of water to sluices. The main use of these excavators in the first two weeks was in-filling behind first aid sandbag defences lines, more especially where breaches had caused deep scours behind the walls.

(Insert here in new paragraph – as marked X overleaf)[27]

[*Inserted text marked X*]

On the river Deben, one of the three wide river estuaries in Suffolk, considerable damage was done to the walls on a river which was able to supply boats. A fleet of 3 motor boats, 2 whalers, 3 pontoons and 2 barges was raised by the A.E.C. and handed over to the River Board for the conveyance of filled sandbags, stakes and baulks of timber. Boats were able to ply in and out of the breaches at high tide, whilst material was dumped on the walls to await landing parties to repair the breaches at low tide. This assistance saved much valuable time in the man handling of sandbags over long distances on the walls. A greater number of motor boats might have been available had it not been for the fact that the majority had had their electrical equipment put out of action by salt water.

In Kent, naval personnel were equipped with landing barges and carried out similar work in the conveyance of sandbags for the repair of forward breaches, where access along the sea walls was impossible owing to the number of breaches awaiting attention and the impracticable walking distance.

[*text on page continues*]

The loan of staff varied directly with circumstances. Local report centres were set up, some by River Boards and others by A.E.C. but in all cases where staff were mixed, there was a complete unity of purpose: each individual tackling a job regardless of who was in charge. In Essex, the main headquarters of the River Board was in Chelmsford, with forward posts at Colchester, Southminster and Grays manned by A.E.C. with Land Agents in charge. The Norfolk and Suffolk River Board had its main centre in Norwich; whilst in Suffolk, sub centres of control at ~~Aldeborourgh~~, Sudbourne, Woodbridge and Ipswich were controlled by Ministry engineers and others loaned by various drainage boards, whilst at Aldeborourgh, a local River Board engineer was in charge. In different circumstances all liason in one area was undertaken by the District Advisory Officer, whilst in another the work was shared with a chairman of a disaster committee. In one instance in Kent, a leading farmer of a district committee took charge.

Broadly speaking, the River Boards through the assistance of Ministry engineers who obtained equipment from all over the country, organised their own responsi-

27 Written very small, underlined and marked with a cross.

bility in the first aid repair of the sea walls. Assistance was also given by A.E.C. in organising labour for this repair. Within the defences, the A.E.C. with assistance from the National Farmer's Union and local authorities supervised the rescue of livestock and the disposal of carcasses: organised emergency grazing for the agistment of stock and the provision of fodders together with special issues of feeding stuff coupons.[28]

Local authorities gave assistance with the loan of surveyors, lorries and small tools. The Womens Voluntary service gave much appreciated help by providing tea canteens almost at the edge of the flood, throughout hard weather.

Whilst the Ministry of Agriculture provided some pumps which were manned by A.E.C. drainage officers a large number were provided by the Admiralty. In addition, the National Fire Service were of great help in providing powerful pumps, in particular at Bawdsey in Suffolk where 9 fire tenders were in operation for 2 days.

[*new page*]

c. The rescue of livestock and the disposal of dead animals

Rescue

The heaviest losses of livestock were in Kent where 765 cattle and 6772 sheep together with pigs and poultry were lost. Rescue operations was probably one of the most strenuous of the emergency operations. At the time of the flood, about 70,000 sheep and 67,000 were on the Kent marshes affected. In the early hours of February 1, stock owners and workers brought thousands of stock to safety, but for many the owners lived as much as 20 miles distant and by the time the news got through, many stock had been lost. One owner lost an entire flock of 700 sheep.

The A.E.C. was helped in the rescue work by local boat clubs and others with boats and in the first few days of the floods, rescued 800–900 bullocks and 7000 sheep.

In Essex where over 1000 sheep were lost, most of the rescue work was carried out by owners. On Foulness Island, which was completely inundated, the Military authorities carried out rescue under the direction of the Ministry and Divisional Veterinary Officer. Stray stock which found their way to safety were subsequently rounded up by A.E.C. staff, yarded and fed by arrangement with farmers and merchants, until they could be drafted back to their owners.

In the East Riding where losses were light, the majority of the animals were borne out to sea by the flood tide.

In Lindsey [*Lincolnshire*], losses were few and rescue was carried out under the direction of owners. Special mention should be made of the work of the cattle transport drivers who moved into the flooded area at dark on the night of the flood. The marsh roads, in common with most in other counties, are bordered by deep ditches and have no boundary marks. The night was black and the flood tide whipped by the gale caused waves which impeded navigation over the flooded roads. The drivers were piloted on the roads by men with local knowledge and their assistance is also commended.

In Norfolk and Suffolk, no rescue work was carried out by the A.E.C. as owners were able to cope with their own situations. In Norfolk, the N.F.U. branch secretaries organised rescue where help was required. In Suffolk, there were several isolated

[28] Agistment being the removal of livestock from one farm to accommodation on another.

cases of particularly commendable rescues of horses, as undoubtedly occurred in other counties.

Disposal of carcases
Where there were large numbers of carcases of dead animals as was the case in Kent and Essex, quick disposal was important and both manure manufacturers and knacker yards were of great assistance in collection. In both Suffolk and Kent, some carcases were burnt using petrol, but on a large scale it was found impracticable and resort was made to mass graves dug by excavators. Where possible, fleeces and hides were removed – before collection by licensed slaughterers. There were also cases of collection of drowned pigs which were sold to a tallow factory.

In many flood areas where the number of carcases were few and where they had been washed up on distant beaches, the police in liason with local health authorities arranged for their disposal. Sanitary inspectors of urban district councils made arrangements for the disposal of drowned domestic animals and carcases of farm stock within the urban boundary. For example, the Felixstowe, Suffolk, U.D.C. arranged for the salvage and burial of 57 pigs, 15 cows, 11 heifers, 1 horse, 27 dogs, 32 cats, 923 fowls and 23 rabbits.[29]

[*further text relocated to intended sequence*]
Continuation from top of page opposite (Disposal of carcases)
There were many instances of difficulty in loading dead animals. The state of the ground for transport running and the 'handling' of heavy wet carcases, together with the inaccessible positions from which dead animals had to be retrieved, made these operations tedious.

[*text on page continues*]
General comment on rescue operations
The work was hazardous. There were many people in crafts of all kinds, who were not familiar either with the water or the craft. Undoubtedly some were unable to swim and others suffered serious illness after the event from exposure in cold. There was a lack of suitable craft both for the loading and carriage of stock and the ideal would be the use of flat bottomed landing craft, both for convenience and safety. In addition, an important factor was the handling of hungry and cold stock in a frightened condition, more especially where they were found in deep water inside a marsh shed.

Note for foot of page[30] (n.b. consideration of stories relating to rescue and salvage operations)

d. The recovery and treatment of flooded farm equipment
At the time of year, very little equipment would have been out in the marsh and the majority of casualties was caused by water entering the farmyard and buildings. Where rescue had to be carried out, owners mainly helped themselves but some local firms of agricultural engineers supplied tackle for rescue work. Many local firms provided a free service for the rehabilitation of machinery. Some moved fleets

29 This example is written small in a tight space at the top of the page.
30 Written at the bottom of the page, separated by a line above.

of service vans into the areas and carried out work on the site, whilst others transported tractors and machinery back to their workshops for treatment.

The best service was probably in the form of advice, whilst the main treatment was the prevention of salt water corrosion. County machine advisors of N.A.A.S. found themselves with a new problem on which many had to seek advice before passing it on to the farmer. This advice included precautions and methods of stripping down, the treatment of metal parts and the avoidance of costly errors such as the running of a tractor with salt water in the sump and bearings.

(Refer to W. Hayles)

e. <u>The organisation of labour for the repair of the sea defences</u>

In order to appreciate the necessity and the reason for the size of the labour organisation and the difficulties of their direction, it is necessary to record a general picture of the situation left by the flood. The technical considerations concerning the evacuation of sea water from the marshes are dealt with in detail elsewhere in the Report and at this juncture it need only be said that it was of vital importance to the land for the salt water to drain off as soon as possible.

The coastal and river estuary marshes are not easily accessible at the best of times. There are few made up roads across the marshes and most are earth tracks which are often impassable for transport in the winter: and in most cases, the track leading away from the farm buildings down to the level of the marshes is no more than an earth road. It was these latter approaches which had to be used, as the whole levels were in flood, and in February their condition is not good. By the time lorries, tractors and trailers and cars had brought workers to the nearest approach to the marsh and other heavy transport conveying equipment, it was only a matter of a few days before these tracks were reduced to a quagmire of deep mud.

The approach to the marshes was one problem whilst the access to the breached walls was quite another matter. The picture was one of a long broken line of defence wall varying form 3–6’ in width at the top, with the sea or a river on one side and the marshes on the other, flooded at high tide to within a few feet of the top of the wall.

The first stage in the repair of the battered walls was the in-filling of breaches with sandbags filled with any soil material which was available and near at hand. Some clay and sand was filled at pits inland and transported, whilst the greater parts for the fillings were taken from soil near the marshes and from saltings mud at low tide. The approach to the problem was a single file procession along the top of the wall and work proceeded from breach to breach. Thousands upon thousands of sandbags were carried by hand along the walls, and where mud was easily accessible at low tide, gangs of men filled bags near the site of the breach, whilst the sealing of the damage was directed by men of the River Boards.

The digging of wet mud whilst standing on a wet slippery surface is not the easiest of jobs at the best of times and in the face of bitter cold winds and snow, for a period often exceeding 7 hours a day, it was not unnatural for workers to get slowed down.

Weather conditions and access to the problem were therefore factors which made it necessary to employ as large a number of men as could be made available, in order to complete the first stage of repairs within a fortnight of the great gale. The reason for this timing was on account of the return of high tides which might again be accompanied by N.W. winds.

As soon as reports of a rough survey of the damage was summarised in London, it was clear that assistance of personnel from the armed forces should be made available immediately. Within a few days, thousands of Naval, Military and R.A.F. forces were posted to the coastline. Where local barracks could be utilized, the transfer was comparatively simple: in other cases, adjutants in charge of troop movements, contacted A.E.C.'s for assistance in finding billets or accommodation in camps formerly occupied by P.O.W.'s farm labour and latterly E.V.W.'s. In many cases, this involved labour officers of the A.E.C. with hurried arrangements for camp equipment with the Ministry of ... [*this line is unfinished*]

Immediately following the disaster members and staff of the River Board met those of the A.E.C. and agreed on the lines of assistance required, and a broad division of responsibility. Both the size of the flood and the damage to sea walls, varied considerably and therefore in some counties, the River Board were able to undertake all of the arrangements for organising labour and leave the A.E.C. to the problems of the farms. Although Kent was faced with a huge task, the Board agreed to be responsible for labour organisation but it had the assistance of A.E.C. staff and members of committee posted in local offices, who helped to direct and arrange the disposition of labour. At the peak, 7000 military were employed, whilst volunteer labour was organised by the N.F.U., industrial concerns, local councils and local farmers. Navy personnel were of great assistance in conveyance of materials to forward breaches.

In Essex, the labour force was organised jointly by the River Board and the A.E.C. The Board being responsible for their own employees, members of the armed forces and labour recruited from the Ministry of Labour. The A.E.C. recruited farmers, farm workers and town volunteers. As a record of this emergency organisation, the following manpower figures relate to Essex and exclude River Board employees.

Farmers and farm workers	4000
School boys	200
Town volunteers	400
A.E.C. staff	160
Naval units	480
Army units	2300
R.A.F. units	2000
County Council	600

The peak of employment was reached on the 13th February when 10,143 were working on the Essex sea defences.

In Suffolk, a similar arrangement was made with the River Board whereby the A.E.C. organised labour from farms, the Ministry of Labour and town volunteers. The first 150 farm workers force eventually grew to 1946 volunteers on the twelfth day of the flood, whilst their numbers were supplemented from many sources, including forestry workers and several thousand members of the armed forces making a total of 3711.

The raising of these workers in Essex and Suffolk was carried out through the medium of the B.B.C., the local press and considerably activity of the N.F.U., including a large number of farmers who acted on their own initiative. The method of liason between the A.E.C. and the River Board for the deployment of workers was identical in Essex and Suffolk. Communication by telephone was open day and night, being transferred from A.E.C. offices to private houses of A.E.C. officials at night. Single offers of labour from one farm or co-ordinated local forces of farm workers were communicated to A.E.C. offices, who in turn informed the H.Q. or local offices of the Board. After deployment of the armed forces, the Board requested further numbers of workers at specific points and in turn the A.E.C.

phoned the farms and directed for the following day. At the close of the day, the engineers in charge of the local offices were informed of the number to expect at operation points on the following day.

The exercise was generally known as Operation Boots, Spades and Grub which farms were asked to supply. Transport was arranged by the A.E.C. in lorries and hired buses, but the greater part of the transport arrangements were left to one farmer in a district to make contacts with his neighbours and arrange any form of transport which was most suitable. Volunteers from towns were asked to report to a centre whilst arrangements were made by the A.E.C. with local bus companies for their transport to and from the sites.

General observations

a. The desirability of a clear cut agreement on labour organisation was shown to be useful. The size of the disaster and the immediate need for action left little time for detailed arrangements and consideration should be given now by both River Boards and A.E.C.s for the drawing up of a plan of organisation. The working party in no way wishes to criticise the arrangements which were made, but emphasises that quicker and smoother arrangements could be effected on other occasion by making use of experience.

b. The deployment of large numbers to strange areas, calls for more detailed arrangements of meeting transports, making up suitable size gangs and directing men to the place of work. Liason between organisers and those responsible for directing the operation could be improved by plans worked up in consultation.

c. Whilst large groups of men on sites, worked remarkably well, the effective output of the work could have been the greater if supervisors of the groups had been appointed and who were able to give technical supervision both to the filling and more especially to the method of laying sandbags. Engineers did give directions in general terms, but their task of survey was so great that it was impossible for them to remain at any one site of work for any length of time. This point could be met by capable supervision of the River Board.

d. The self employment of volunteer enthusiasts is to be discouraged. Much valuable time was lost by work being carried out by these people, all of which had to be undone owing to its ineffectiveness against tides.

The employment of unemployed men from the Ministry of Labour was in most cases a travesty of work. In Suffolk, there were complaints of pay, cold and the type of work; there were also agitators who encouraged men to leave the job. The majority of these men were quickly discharged and it is unlikely that many of this class of labour are worth considering on similar occasions.

[*new page*]

Section V. (d.) (i.)

The pre-flood state of marsh drainage

Essex The pre-flood condition of the marsh ditches were in general, in good order. A very large proportion of the marsh arable area is still ploughed on the stetch: in some cases, it is on both stetch and bed.[31] Again a large proportion of the grassland was laid down on the stetch. In addition, the arable is extensively water

[31] A stetch is a ridge between two furrows in ploughed land.

furrowed. This method of cultivation to provide surface drainage remains necessarily where most of the land is an alluvium on an impervious London clay. The success of this farming can only be achieved if ditches are maintained to convey this surface water.

Some of the land is pipe drained and some with pipe cum mole and in more recent years with the introduction of pumping stations on some levels, the effectiveness of under drainage has aided the maintenance of these systems which were on the whole in good order at the time of the flood.[32]

Kent The ditches in the arable areas where the soil is alluvium over clay as is found in the north west and in the variable soils of the east of the county were well maintained. On the other hand, the ditches on the 32,000 acres of flooded grassland were for the most part in poor order. Many were silted up for lack of maintenance and very little mole draining had been carried out although the land was suitable. It must also be added that the condition of many of the main ditches and the coastal sluices were poor and no doubt contributed to the general state of affairs.

The areas of heavy London clays with poor grassland and a dense thick matt were, as elsewhere, associated with poor drainage but on similar heavy soils such as are found in the Rochester, Sheppey and Faversham districts, where drainage and general management is above average, the state of the grassland was very different and the better drainage conditions gave quicker response to flood recovery.

Lincolnshire (Holland) The flood covered an area of light sandy loam which is highly fertile and has for long been farmed on a high standard and which has necessitated the maintenance of drainage systems, all of which were in good order at the time of the floods.

Northumberland The soils of the flooded areas ranged from medium heavy to heavy silts changing to sandy silts near the sea. Most of this land is tile drained and these systems and the ditches were in good condition. The flood rough grazing not under drained.

Yorkshire (East Riding) There are some old pipe drain systems in the Holderness area with some recent mole drainage along the north bank of the river Humber. The ditches were maintained in fair order.

Lincolnshire (Lindsey)
Add note.
[*a sheet of paper has been pinned to the page, with the following text*]
The efficiency of drainage was good on the arable and poor on the grassland. Most of the arable is tiled drainage with both ditches and main drains kept in good order; in particular the better quality silts, maintained a high standard of drainage. The ditches over the grassland areas were not well maintained, largely on account of the traditional system of retaining a high level of water in the summer for stock drinking. The soil of the grassland was capable of mole draining but no work had been carried out.

32 Mole drains, a sub-surface drainage system of unlined channels formed in (typically clay) subsoil. They do not drain surface water but lower ground water levels below the root zone of the crop.

[*text in notebook continues*]

East Suffolk The marshes here have variable soils and the general state of the ditches were fair. The marshes in the Waveney Valley have varying depths of sandy peat with shallow surface drains which are difficult to maintain. The levels are pumped. Further south to centre of the coastline, the soil is a silty clay loam with varying depth of surface peat. South of the river Alde the soils are heavier and for the most part are clay loams: ditches at the time of the floods were good in places and bad in others.

As in Kent, the areas of poor grassland were all associated with poor matted grassland which received little or no management and these were found all down the coast, regardless of soil. In areas where the heavier productive soils from Aldeburgh to Felixstowe had in the past 10 years been ploughed, ditches were in good order, whilst those adjacent to the grass in the same areas were as neglected as the grass.

Prior to the flood many sluices were in a poor state of repair and their ability to discharge marsh water had the effect of retaining a high ditch water level which in turn was a deterrent to better farming. There were a few privately operated pumps. In consequence no marsh levels had been under drained for a great many years. The only existing pipe drains cover a few acres on the north bank of the Stour.

Norfolk Also a county of variable soils and equally variable conditions of drainage prior to the flood. In the west of the county, the Downham area is pumped: soil is a poor peat, there is no under drainage and the ditches need enlarging. In the Watlington, Magdalen and Terrington areas which are pumped, the ditches were good and much of the area piped drained. In the Lynn–Hunstanton, also pumped, some of the arable is piped, but the ditches of the area were only fair and needing maintenance.

In the north where water exit is by wall sluices, the soils are heavy silt loams to heavy clays, with no under drainage. The Hunstanton–Holkham area ditches were poor and not maintained; likewise the ditches across to Wells, Stiffkey and Cley. From Salthouse to the River Glaven the ditches were in fair order. In the east, the ditches of the sand and peat of East Palling were fair; whilst the condition of water-courses on the peat over clay marshes of Acle and Haddiscoe was variable. Inspite of pumps, the summer grazing systems involved the retention of high ditch levels for stock watering. None of these areas in the north or east are under drained.

In the Bloefield and Flegg districts the marshes are a varying depth of peat over clay. Some is under drained and systems were maintained as the standard but where there was no under drainage, the maintenance was poor and lettings for summer grazing permitted a high water level in the ditches.

(ii.) Drainage problems left by the flood

The aftermath of the flood left problems common to all counties concerned and certain features in relation to soil type and drainage conditions prevailing at the time of the flood.

Where breaches of sea and river walls occurred, the clay embankment burst into the marsh and blocked a section of the delph ditch and varying lengths of ditches leading down to the delph. This was not the case with all wall breaks as many burst through interior pressure of the flood and the embankment burst outwards to sea.

Flood borne debris of tons of reeds and spartina grass, tree logs and large sections of soil and decomposed organic matter torn from the sides of ditches and fleets were blocking ditches over distances of many chains in length.

Sand and shingle washed from the beaches was borne by the tide over the walls and sand dunes and covered extensive areas of marsh, also blocking the ditches to the delph and in certain cases blocking the mouths of main rivers. In consequence, the debris caused blocking of outlet pipes in the ditches.

On levels where the soil was light, especially deep sandy peat type deposits, there were cases of damage to the sides of the ditches which slipped under the weight of water and debris.

(iii.) Action of county staffs, following the flood

Before any attention could be given to the restoration of drainage by drainage departments of the A.E.C., it was necessary to allow time for the marshes to dry to a reasonable state after the flood water had drained off via the sluices. As soon as it was possible, attention was given to the release of water at the outlet. Whilst the River Boards attended to temporary repairs to sluices, both they and the A.E.C.s excavators cleared the delph ditches and the lengths of marsh ditches which led down to the delph, of flood borne debris.

Most of the trouble preventing the free flow of water was caused in the vacinity of the wall breaches. Here the walls burst into the marsh, hurling tons of clay into the delph and into the end sections of the connecting ditches. Where the foreshore was a sand and shingle beach, the tide bore thousands of tons of the beach over the wall on to the marshes. Where breaches occurred, the beach material poured in to fill ditches and often covering several acres of the marsh. In addition, ditches were filled with flood borne debris more especially in the areas of tidal estuaries.

The immediate task was the release of the remaining salt water in the ditches and the run off from pools which stood on the marshes. No importance was attached to the 'finish' of the excavation work which forged ahead as fast as possible with the purpose of dragging the debris clear to allow water to run. No attempt was made to leave the ditch in the normal state with a finished batter on the banks. Both the supply of equipment, (see top of page) [*squeezed in at the top of the page in very small writing, in different ink*] mainly excavators and bulldozers and also available staff, were in several counties, entirely inadequate to cope with the situation. Arrangements were made for drainage officers to be loaned from inland counties.

(iv.) Flood drainage grants

For this immediate work on mains drains and farm ditches which were blocked by flood borne debris, the full cost of the work was reimbursed provided the work was completed by July 31, 1953. The subsequent attention required on all ditches on the marsh to ensure a free flow of the salt water and the subsequent dispersal of salt laden water after rains, the prevailing rate of 50% grant was raised to 75%. In the first place the work should have been completed by December 31, 1953, but owing to the immesurable task, the closing date was extended on 3 further occasions to finally close on 31 October, 1953.[33] This increased grant also extended to the provisions for new tile and mole drainage systems.

The criterion of eligibility for attention to ditches was the state of the ditch to provide an unimpeded flow, rather than any actual damage done to the ditch. The removal of silt, reeds and rushes by excavation increased the flow and released land pipes.

33 It eventually closed in October 1955, not 1953.

Section V. (c.) Coastal Flooding (Emergency Provisions) Act 1953
The object of the act was to make provision for the restoration of the flooded land and damage to sea defences so that the areas concerned could resume their role in food production.

Part I. of the Act gave power to the Ministry of Agriculture to authorise River Boards to enter upon land to construct new walls and to take clay and other materials to build access walls and to take land for essential purposes as may be required. Clauses were inserted to enable consideration to be given to compensation in cases where there was injury affected by reason of the works done.

(i.) Acreage payments to farms and allotments
Part II. of the Act enabled the Ministry to make schemes over the period 1953–57 inclusive for acreage payments to farmers with a view to securing the rehabilitation of farmland which in consequence of the sea floods had been rendered unfit for the full production of crops. The rates of payment per acre varied from year to year and according to whether the land was in crop, bare, laid down or remained in grass. Payements were made provided an agreed programme of rehabilitation was entered into by the farmer and the County Agricultural Executive Committee. The scheme also included allotments aggregating over half an acre, but not to private gardens.

(ii) Other rehabilitation payments
The flood caused very considerable damage to fixed equipment and left problems in the form of tidal borne debris. The Act also provided for County A.E.C.s to undertake reinstatement, or arrange to supply contractors or reimburse the occupier for the approved cost of work carried out with farm labour.

The undertakings included, the cost of grubbing orchards, the removal of salt laden glasshouse soils, the repair and reinstatement of fences, hedges, farm bridges and roads, and the removal of debris, sand and shingle. The certification of these records was undertaken by A.E.C. staff and payments were made up to a time when it was decided to transfer the liability to the Lord Mayors Flood and Tempest Distress Fund. (See vi. (2) f. vii.)

(iii.) Acreage payment awards
Year 1953
Payments where crops were in the ground on 31 January, 1953.

Arable	£ per acre
Asparagus, bulbs & nursery stock inc. flowers	80
Strawberries	60
Winter brocolli, spring cabbage, winter lettuce	40
Other annual horticulture crops	30
Wheat, barley, oats, rye, dredge corn, feeding beans, flax, linseed[34]	20
Row crops grown for seed	25
Other commercial crops	15

34 Dredge corn, a mixed crop of oats and barley used for stock feed.

Grassland	£ per acre
Temporary grass, wholly or mainly destroyed and unusable in 1953	10
Permanent grass, [ditto]	6
Temporary grass, partly destroyed but capable of some use in 1953	7
Permanent grass, [ditto]	4
Approved crops sown on bare land after 1 January, 1953	Half arable land rates
Bare land not to be used for a crop or grass in 1953	8

Year 1954

Where land was on 31 January 1953 under the following crops and the retention thereof uncropped was approved, payments were made as follows:

	£ per acre
Asparagus, bulbs & nursery stock inc. flowers	40
Strawberries	30

Where land was pasture on 1 January 1954 and is reatined and managed in an approved manner, payments as follows:

	£ per acre
Grass which is wholly unusable in 1954	8
Grass capable some use in 1954	5

Where the use of the land for the growing of any of the crops specified below has been approved, payments were as follows:

	£ per acre
Horticultural crops	17½
Wheat, barley, oat, rye, dredge corn, and row crops grown for seed	12
Feeding beans, flax, linseed, sugar beet and row and fodder crops (except grass)	10
Other commercial crops (except grass)	7½

Where the land is bare January 1, 1954 and the retention thereof is approved, the payment was at £8 per acre.

Year 1955

	£ per acre
Land which was in pasture on 1 January, 1955 and which was retained as pasture	6
Land which is used for the growing of crops or for the laying down of pasture	10
Land which is bare on 1 January, 1955 and which is retained uncropped	10

APPENDIX

COMPARABLE LOSSES TO AGRICULTURE IN ADJACENT COUNTIES

Agricultural lands within all the counties flooded by the sea surge on 31 January–1 February 1953 suffered losses and damages to varying degrees. In order to give context to the figures Trist recorded for East Suffolk, the official statistics for East Riding of Yorkshire, Lincolnshire, Norfolk, Essex and Kent are summarised below.[1]

Losses and damages to agricultural land and farm stocks, for six counties affected by sea floods in 1953

	Yorks. E. Riding	**Lincs.**	**Norfolk**	**Suffolk**	**Essex**	**Kent**
	Number of holdings of one acre and over affected by the flood					
	100	596	828	251	328	321
	Total area of agricultural land flooded, and its use					
Arable	2,706	14,383	9,670	3,975	15,075	5,035
Grass	1,797	12,094	21,984	16,475	23,711	32,661
TOTAL	4,503	26,487	31,654	20,450	38,785	37,695
	Total area of agricultural land still flooded after 21 days					
	169	291	unknown	5,147	11,000	9,300
	Livestock losses					
Cattle	29	142	114	258	259	870
Horses	–	–	15	11	6	4
Pigs	128	709	1,220	86	478	598
Poultry	2,569	18,475	19,656	1,502	7,177	3,563
Sheep	333	214	613	303	1,008	6,774

1 These figures are from the appendix of the official MAFF report to which Trist contributed. Additional statistics for losses by crop type, for farm stocks and equipment, and within horticulture are available in MAF 221/11, 'The Effect on Agriculture of the East Coast Floods, 1953' [Draft report of the working party set up by the main advisory committee on sea flooded land, MAF]. See also pp. 32, 88–90 above.

BIBLIOGRAPHY

Unpublished primary sources

The National Archives, Kew

MAF 157/23, County Agricultural Relief Committee: Correspondence with Lord Mayor's National Flood and Tempest Distress Fund

MAF 157/25, Reports and Surveys on Extent of Land Flooded in East Suffolk

MAF 220/9, Correspondence between East Suffolk AEC and MAF Floods Emergency Division, Whitehall

MAF 221/11, 'The Effect on Agriculture of the East Coast Floods, 1953' [Draft report of the working party set up by the main advisory committee on sea flooded land, Ministry of Agriculture and Fisheries]

Suffolk Archives, Ipswich

A2727/1/1, 'The Great Sea Floods of 1953' and 'Records in the History of the Suffolk Coastline', typescript draft monograph by Trist

A2727/1/2, 'The Sea Floods 1953 in Suffolk', diary of observations by Trist

A2727/1/3, 'Sea Floods 1953 – Report', notes on action taken by Trist

A2727/1/4, Records of coastal flooding, East Suffolk (1915–70)

A2727/1/6, Coastal flooding and sea defences, press cuttings (1950–90)

A2727/1/7, 'The Effect on Agriculture of the East Coast Floods, 1953', NAAS report (draft)

A2727/1/8, Salinity and reclamation of soils, notes and pamphlets

A2727/1/11, 40th anniversary commemorative coverage as cuttings (1993)

A2727/2/1–16, Maps of Suffolk coastal lands inundated by 1953 floods

K681/2/93/3, Photograph of Shottisham Creek taken in February 1953 by Robert Adams, chief engineer for the Isle of Wight River Board

GC462/3/1/1, NFU, Suffolk County Branch, members' year books and annual reports, 1945–65

HA412, Rope Family Archives (see also HA28, HA81and HA444)

HD1848/2, Booklet produced by the Women's Voluntary Service concerning flood relief work, 'Report on Help Given by East Suffolk W.V.S. after the Storm and Tempest on January 31st–February 1st 1953'

Printed primary sources

BPP 1953–54, XIII.511, 'Home Office, Scottish Office, Ministry of Housing and Local Government, Ministry of Agriculture and Fisheries, Report of the Departmental Committee on Coastal Flooding' [Waverley report]

Grieve, H., *The Great Tide: The Story of the 1953 Flood Disaster in Essex* (Chelmsford, 1959)

Institution of Civil Engineers, *Conference on the North Sea Floods of 31 January/1 February 1953: a Collection of Papers Presented at the Institution in December 1953* (London, 1954)

Lord Mayor of London's National Flood and Tempest Fund, *The Sea Came In: The History of the Lord Mayor of London's National Flood & Tempest Distress Fund* (London, 1959)

Newspapers and periodicals

Agriculture, The Journal of MAF
East Anglian Daily Times (EADT)
Kelly's Directory of Felixstowe and Neighbourhood
Kelly's Directory of Ipswich and Neighbourhood
Suffolk Farmers' Journal
Suffolk Magazine
Supplement to The London Gazette
The Guardian
The Royal Engineers Journal

Secondary sources

Armstrong, S., and Easthope, L., '"That Night": Unlocking the Memories of Loss on Canvey Island in 1953', *International Journal of Regional and Local History* 13, No. 2 (2018), pp. 134–46

Baxter, P.J., 'The East Coast Big Flood, 31 January–1 February 1953: A Summary of the Human Disaster', *Philosophical Transactions of The Royal Society A,* No. 363 (2005), pp. 1293–312

Brassley, P., Harvey, D., Lobley, M., and Winter, M., *The Real Agricultural Revolution; The Transformation of British Farming 1939–1985* (Woodbridge, 2021)

Cocroft, W., and Alexander, M., *Atomic Weapons Research Establishment, Orford Ness*, English Heritage Research Department Report 10 (2009)

Cole, W., *A Poetical Sketch of the Norwich & Lowestoft Navigation Works* (Norwich, 1833)

Copping, A.C., 'Philip John Owen Trist OBE, BA, MRAC, FLS 1908–1996', *Suffolk Natural History* 33 (1997), pp. 100–01

Cotton, K.E., 'Flood Damage in Norfolk and Suffolk', *Conference on the North Sea Floods of 31 January/1 February, 1953: a Collection of Papers Presented at the Institution in December 1953* (London, 1954), pp. 200–11

Flaxman, R., *Wall of Water: Lowestoft and Oulton Broad During the 1953 Flood* (Lowestoft, 1993)

Hall, A., 'The Rise of Blame and Recreancy in the United Kingdom: A Cultural, Political and Scientific Autopsy of the North Sea Flood of 1953', *Environment and History* 17, No. 3, (2011), pp. 379–408

Liddiard, R., and Sims, D., *A Very Dangerous Locality: The Landscape of the Suffolk Sandlings in the Second World War* (Hatfield, 2018)

Macpherson, J., *The Felixstowe Floods of 1953: Never to be Forgotten* (privately published, 2023)

O'Hara, G., *The Politics of Water in Post-war Britain* (London, 2017)

Pollard, M., *North Sea Surge: The Story of the East Coast Floods of 1953* (Lavenham, 1978)

Rennoldson-Smith, P., *The 1953 Essex Flood Disaster: The People's Story* (Cheltenham, 2012)

Steers, J.A., 'The Suffolk Shore, Parts I, II and III', *Proceedings of Suffolk Institute of Archaeology and Natural History* 19 (1927)

Steers, J.A.,'The East Coast Floods', *The Geographical Journal* 119, No. 3 (1953), pp. 280–95

Storey, N.R., *Flood Alert! Norfolk 1953* (Stroud, 2003)

Summers, D., *The East Coast Floods* (Newton Abbot, 1978)

Trist, P.J.O., *A Survey of the Agriculture of Suffolk* (London, 1971)

Trist, P.J.O., *Land Reclamation* (London, 1948)

Trist, P.J.O., 'Ecology in the Reclamation of Salt flooded Marshes', *Agriculture* 59, No. 12, March 1953, pp. 571–3

Trist, P.J.O., 'Suffolk Marsh Reclamation Policy', *Agriculture* 61, No. 7, November 1954, pp. 328–32

Trist, Richard, 'Bibliography of P.J.O. Trist , OBE, BA, MRAC, FLS, Agriculturalist and Botanist (1908–1996)', *Suffolk Natural History* 39 (2003), pp. 75–84

Waldringfield History Group, *Waldringfield* (Chelmsford, 2020)

Wells, D., 'John Trist (1908–1996)', *Nature in Cambridgeshire*, No. 39 (1997), p. 89

Broadcast media

'Learning From the Great Tide', BBC Radio 4, first broadcast on Monday 30 January 2023

'The Long View: The Big Flood of 1953', BBC Radio 4, first broadcast on Tuesday 28 January 2014

Websites

Abbott's Hall Almanac blog from the Food Museum: https://abbotshall.wordpress.com

BBC Eye witness accounts recorded in 2003: www.bbc.co.uk/suffolk/dont_miss/floods/eye_witness_accounts/bernard_adams_fx.shtml

Essex Record Office blog: www.essexrecordofficeblog.co.uk/the-great-tide-remembered/#:~:text="The%20Great%20Tide"%20was%20written,complete%20story"%20of%20the%20disaster

History of 1952 Lynmouth Flood Disaster from Visit Lynton & Lynmouth website: https://visitlyntonandlynmouth.com/history-heritage/the-1952-lynmouth-flood-disaster/

Memories of Noreen Prichard Carr from her daughter's website: https://retro-cooking.co.uk/about/

Suffolk Food Hall website: https://suffolkfoodhall.co.uk/food-nall-through-the-ages/

Southwold Museum website: www.southwoldmuseum.org/thesea_1953Floods.htm

MAPS OF SUFFOLK COASTAL LANDS
ANNOTATED BY P.J.O. TRIST

1. A2727/2/1, Brantham–Harkstead (sheet 62/13)

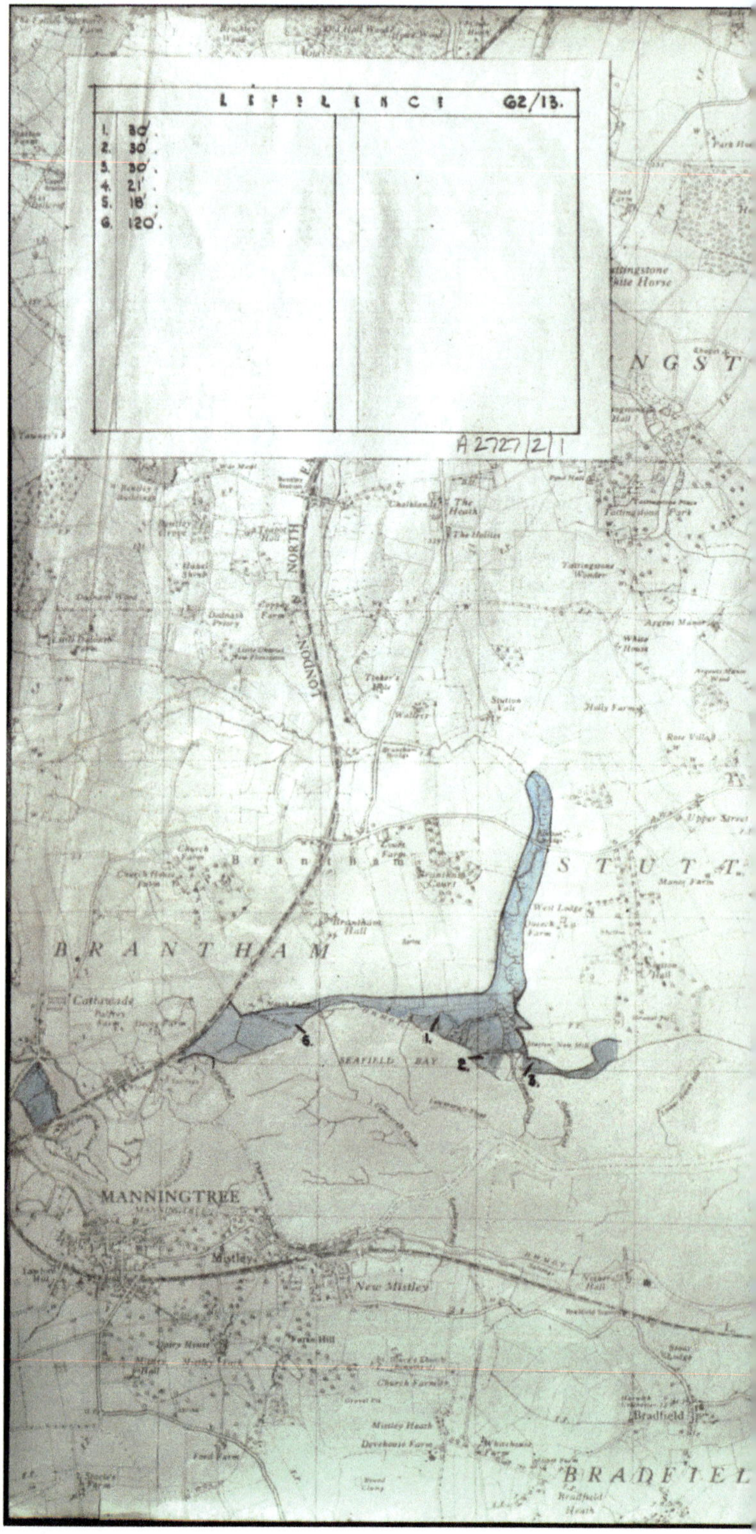

RIVER ORWELL
Freston
Woolverstone
FRESTON
WOOLVERSTONE
Woolverstone Hall
Turkey Farm
Grove Farm
Brown's Farm
Brook Farm
Holbrook
Holbrook Gardens
HOLBROOK
Holbrook House
HARKSTEAD
Lower Holbrook
Harkstead
Alton Green Farm
Nether Hall
Sparrow Hall
5.
4.
HOLBROOK BAY
St. Clement's Church
THE SEA FLOOD
JANUARY 31ST – FEBRUARY 1ST 1953.
MINISTRY OF AGRICULTURE AND FISHERIES.
CROWN COPYRIGHT RESERVED.
PRESENTED BY THE EAST SUFFOLK AGRICULTURAL EXECUTIVE COMMITTEE TO THE IPSWICH AND EAST SUFFOLK RECORD OFFICE.
MARCH 6 1954
COUNTY AGRICULTURAL OFFICER.

2. A2727/2/2, Ipswich (sheet 62/14)

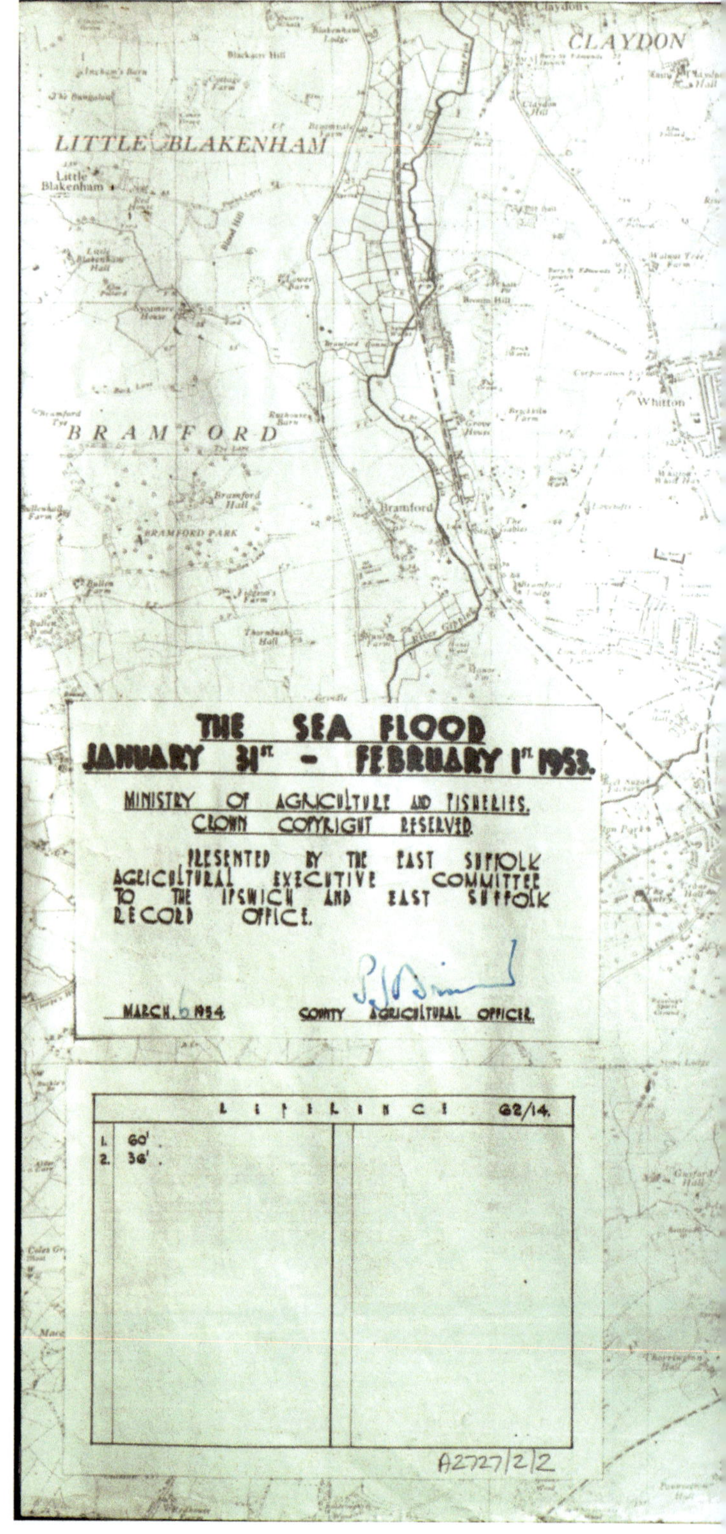

TUDDENHAM
Tuddenham
WESTERFIELD
Westerfield
IPSWICH
Rushmere St. Andrew
Priory Heath
Gainsborough Estate
Rose Hill
Wherstead
WHERSTEAD PARK
STOKE PARK
2.
1.

3. A2727/2/3, Erwarton–Felixstowe (sheet 62/23)

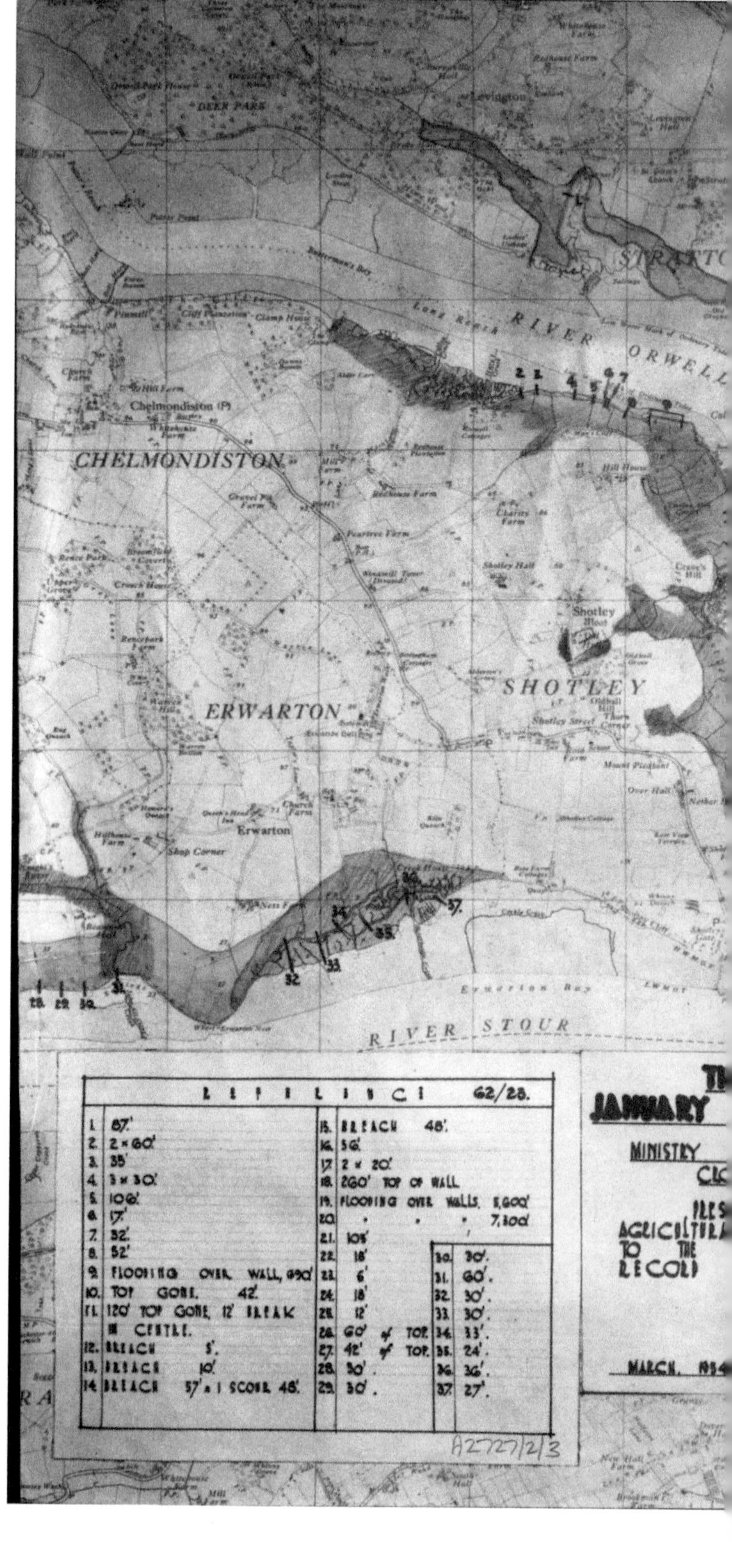

Kirton
Falkenham
TRIMLEY ST. MARTIN
Trimley Heath
Trimley
TRIMLEY ST. MARY
Walton
Waltonferry
Landguard Marsh
ORWELL HAVEN
HARWICH
HARBOUR
LANDGUARD COMMON
Landguard Point
12
13
15
16
17
18
19
20
22
23
24
EA FLOOD
FEBRUARY 1ST 1953.
CULTURE AND FISHERIES.
RIGHT RESERVED
THE EAST SUFFOLK
UTIVE COMMITTEE
AND EAST SUFFOLK
UNTY AGRICULTURAL OFFICER.

4. A2727/2/4, Kirton–Woodbridge (sheet 62/24)

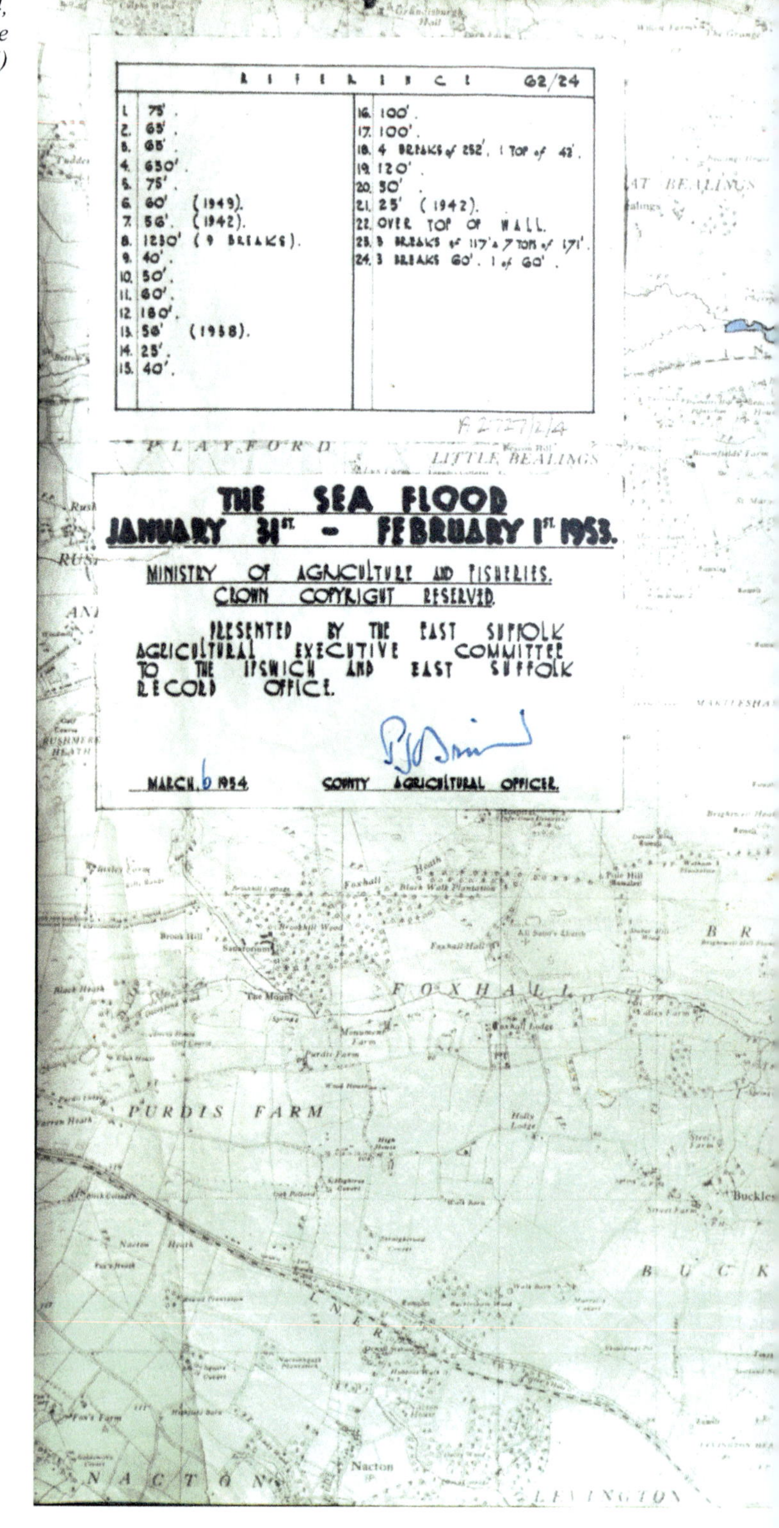

WOODBRIDGE
WALDRINGFIELD
Waldringfield
Newbourn
Hemley
Reddenpit Hill
Sluice Farm
Corporation Farm
Manor House
Tet Hill
Sutton Walks
Little Haddon Hall
Hill Farm
Heath Farm
Street Farm
Hemley House
Kirton Lodge
Kirton Marshes
Church Farm
Stonner Point
1.
2.
3.
4.
5.
6.
7.
8.
9.
10.
11.
12.
13.
14.
15.
16.
17.
18.
19.
20.
21.
22.
23.
24.

5. A2727/2/5,
Felixstowe–Bawdsey
(sheet 62/33)

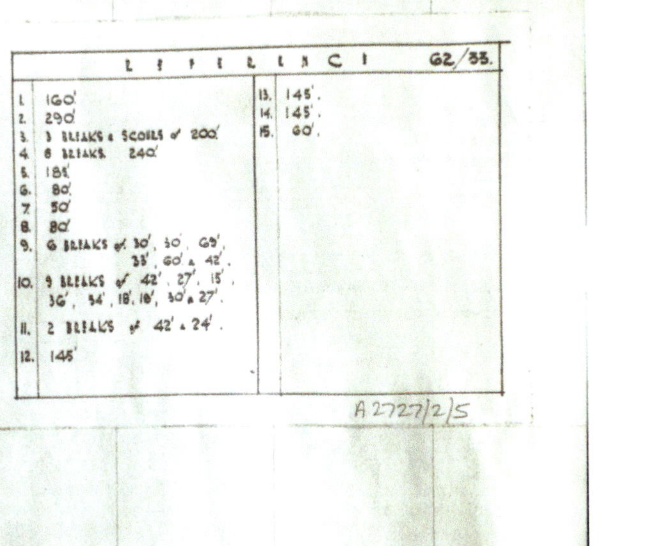

REFERENCE 62/33.

No.		No.	
1.	160'	13.	145'.
2.	290'	14.	145'.
3.	3 BREAKS & SCOURS of 200'	15.	60'.
4.	8 BREAKS 240'		
5.	185'		
6.	80'		
7.	50'		
8.	80'		
9.	6 BREAKS of 30', 30', 65', 35', 60' & 42'.		
10.	9 BREAKS of 42', 27', 15', 36', 34', 18', 18', 30' & 27'.		
11.	2 BREAKS of 42' & 24'.		
12.	145'		

A 2727/2/5

THE SEA FLOOD JANUARY 31ST – FEBRUARY 1ST 1953.

MINISTRY OF AGRICULTURE AND FISHERIES.

PRESENTED BY THE EAST SUFFOLK AGRICULTURAL EXECUTIVE COMMITTEE TO THE IPSWICH AND EAST SUFFOLK RECORD OFFICE.

MARCH 6 1954 COUNTY AGRICULTURAL OFFICER.

6. A2727/2/6, Ramsholt–Butley (sheet 62/34)

ST. ANDREW
Capel St. Andrew
Scotland Heath
Butley Abbey
Butley Corner
Butley Marshes
Barrow Hill
Duck Farm
Boyton
BOYTON
Boyton Hall
Boyton Marshes
Cauldwell Hall
Duck Corner
Spring Farm
Church Farm
RIVER ORE
HOLLESLEY BAY
ORFORD HAVEN
1.
2.
3.
7.
8.
9
19.
THE SEA FLOOD
JANUARY 31ST - FEBRUARY 1ST 1953.
MINISTRY OF AGRICULTURE AND FISHERIES.
CROWN COPYRIGHT RESERVED.
PRESENTED BY THE EAST SUFFOLK AGRICULTURAL EXECUTIVE COMMITTEE TO THE IPSWICH AND EAST SUFFOLK RECORD OFFICE.
MARCH. 6 1954.
COUNTY AGRICULTURAL OFFICER.

7. A2727/2/7, Butley–Snape (sheet 62/35)

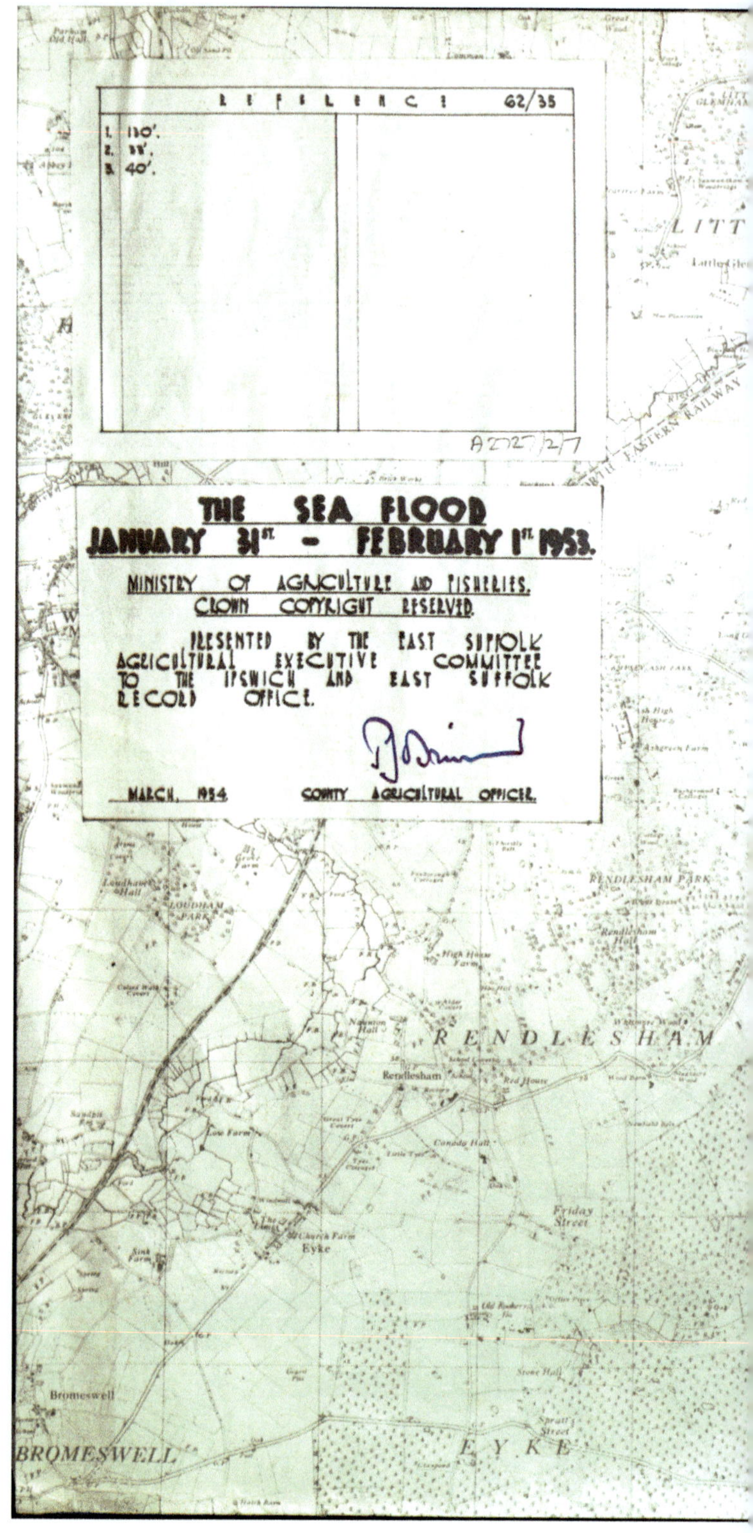

FARNHAM
SNAPE
EMHAM
Blaxhall
XHALL
TUNSTALL
TUNSTALL FOREST
Tunstall
Tunstall Common
Kimple Hill
CHILLESFORD
Chillesford
WANTISDEN
Wantisden Heath
Butley
BUTLEY

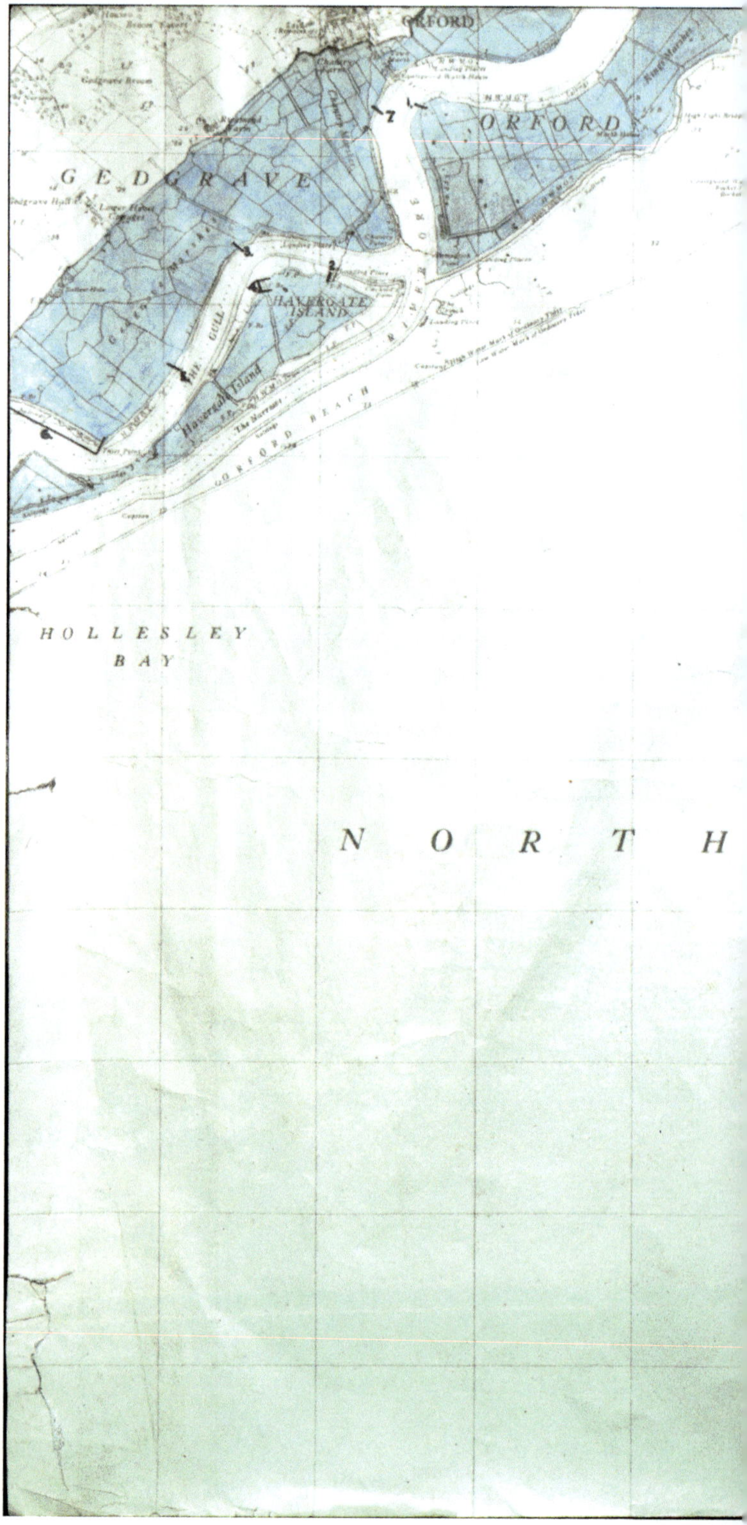
ORFORD
G E D G R A V E
O R F O R D
HAVERGATE ISLAND
Havergate Island
THE GULL
R I V E R   O R E
O R F O R D   B E A C H
H O L L E S L E Y
B A Y
N O R T H

8. A2727/2/8, Gedgrave–Orford (sheet 62/44)

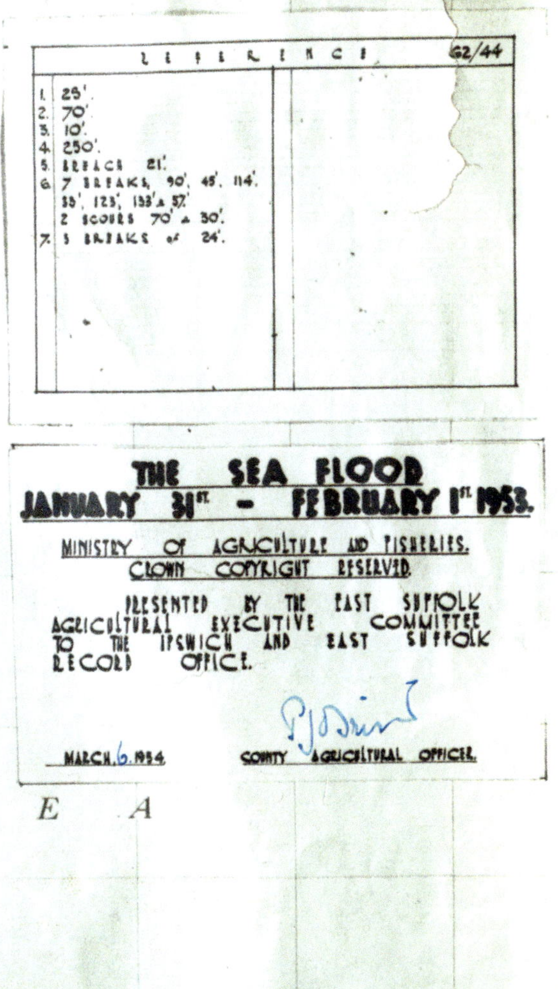
ORFORD NESS
REFERENCE
62/44
1. 25'.
2. 70'.
3. 10'.
4. 250'.
5. BREACH 21'.
6. 7 BREAKS, 90', 45', 114', 35', 123', 133' & 57'
2 SCOURS 70' & 30'.
7. 3 BREAKS of 24'.
THE SEA FLOOD
JANUARY 31ST. – FEBRUARY 1ST. 1953.
MINISTRY OF AGRICULTURE AND FISHERIES.
CROWN COPYRIGHT RESERVED.
PRESENTED BY THE EAST SUFFOLK AGRICULTURAL EXECUTIVE COMMITTEE TO THE IPSWICH AND EAST SUFFOLK RECORD OFFICE.
MARCH. 6. 1954
COUNTY AGRICULTURAL OFFICER.
S E A

9. A2727/2/9,
Sudbourne–Aldringham
(sheet 62/45)

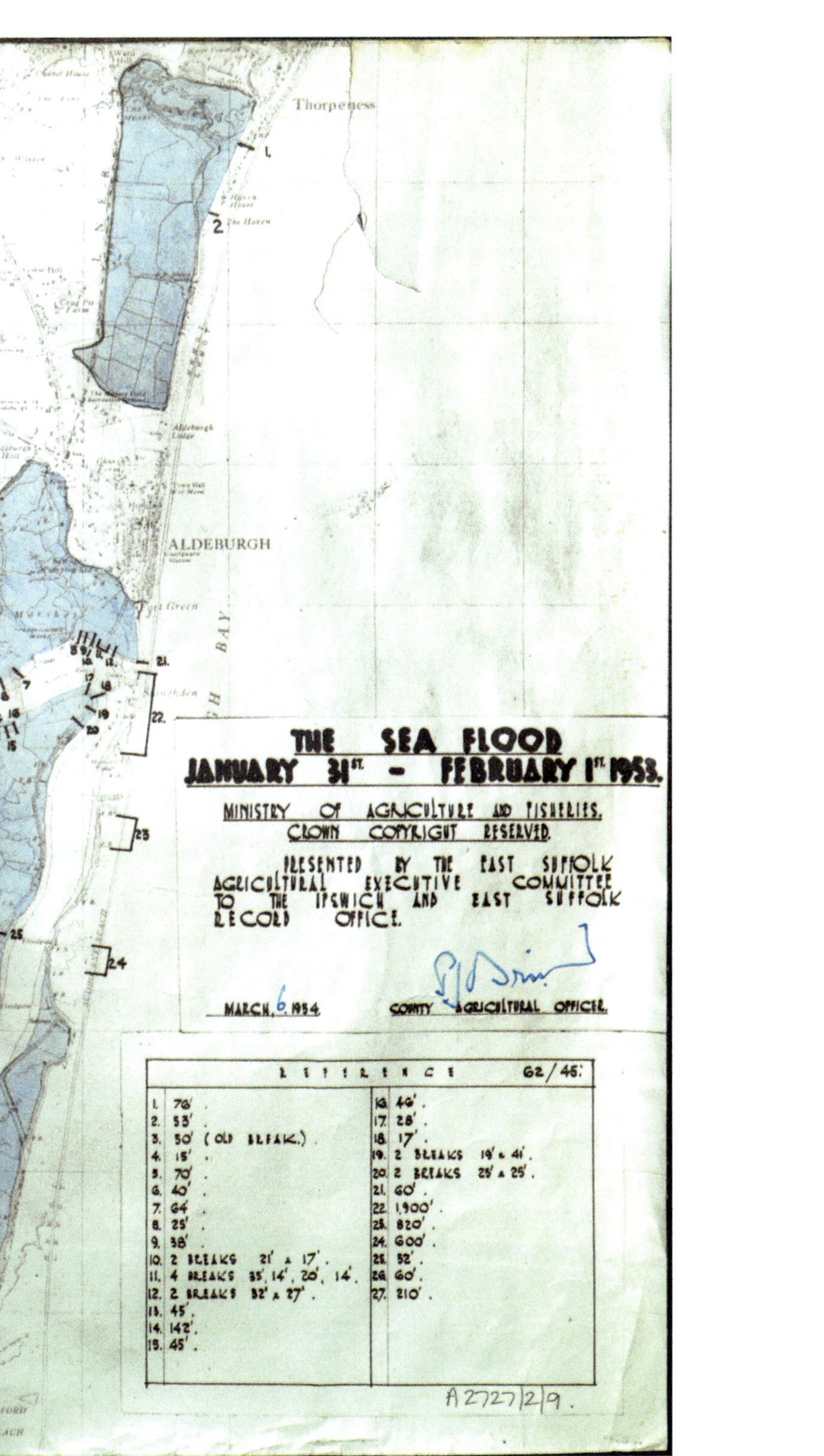

Thorpeness
The Haven
ALDEBURGH
BAY
THE SEA FLOOD
JANUARY 31ST – FEBRUARY 1ST 1953.
MINISTRY OF AGRICULTURE AND FISHERIES.
CROWN COPYRIGHT RESERVED.
PRESENTED BY THE EAST SUFFOLK AGRICULTURAL EXECUTIVE COMMITTEE TO THE IPSWICH AND EAST SUFFOLK RECORD OFFICE.
MARCH, 6. 1954
COUNTY AGRICULTURAL OFFICER.
REFERENCE 62/45.
1. 76'.
2. 53'.
3. 50' (OLD BREAK.).
4. 15'.
5. 70'.
6. 40'.
7. 64.
8. 25'.
9. 38'.
10. 2 BREAKS 21' & 17'.
11. 4 BREAKS 35', 14', 20', 14'.
12. 2 BREAKS 32' & 27'.
13. 45'.
14. 142'.
15. 45'.
16. 46'.
17. 28'.
18. 17'.
19. 2 BREAKS 19' & 41'.
20. 2 BREAKS 25' & 25'.
21. 60'.
22. 1,900'.
23. 820'.
24. 600'.
25. 52'.
26. 60'.
27. 210'.
A2727/2/9.

10. *A2727/2/10, Aldringham–Dunwich (sheet 62/46)*

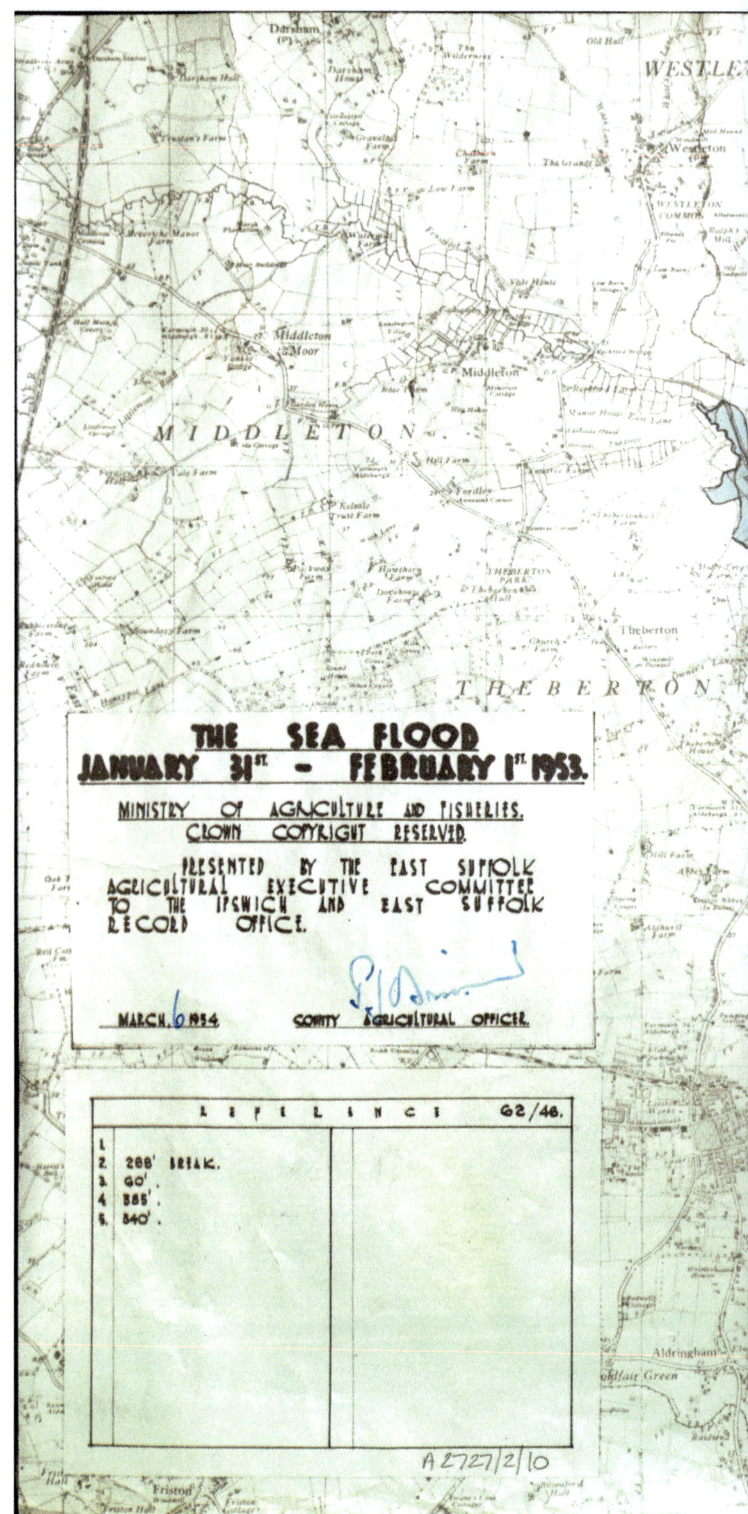

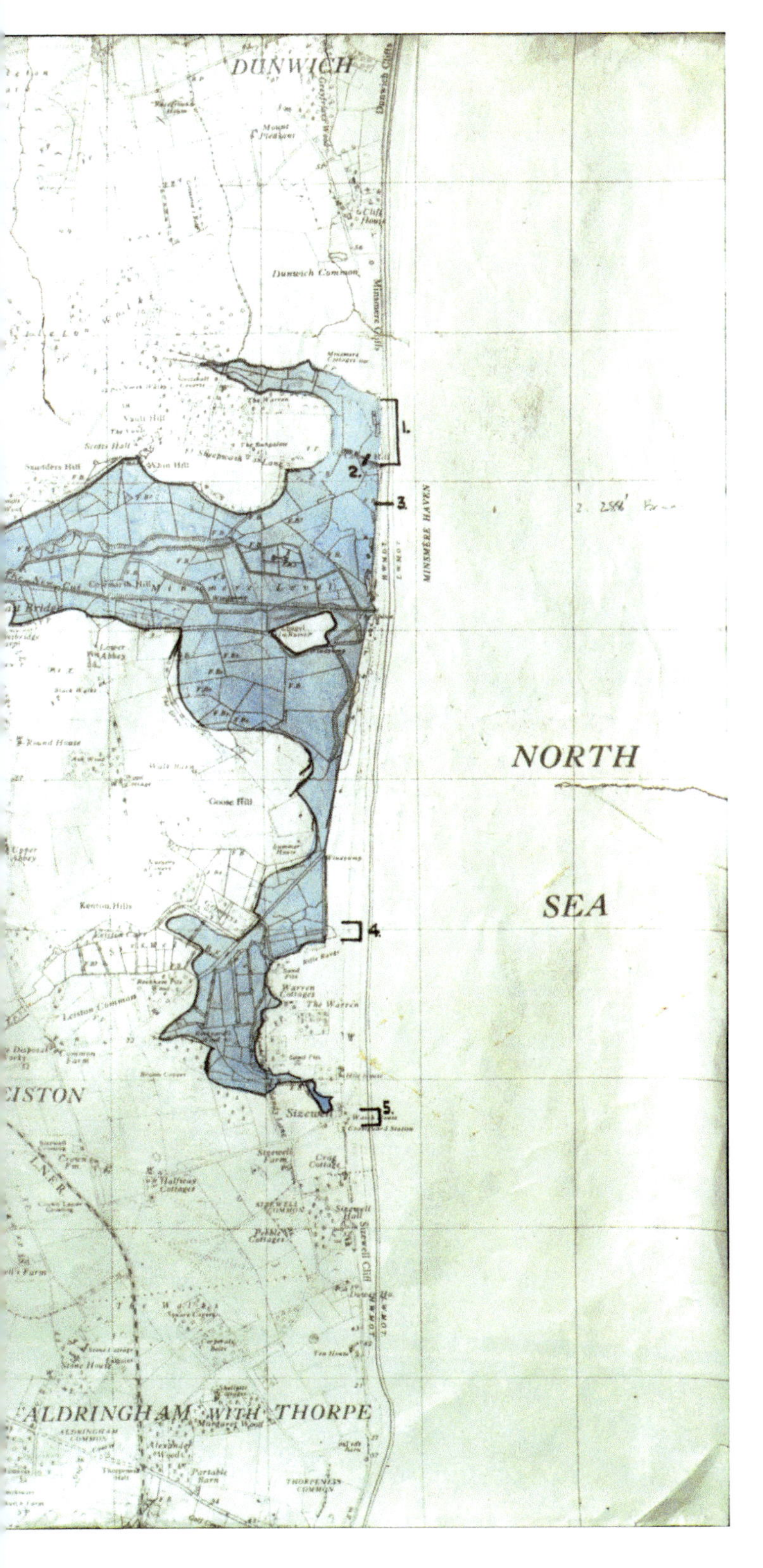
DUNWICH
Dunwich Common
Minsmere Cliffs
1.
2.
3.
MINSMERE HAVEN
NORTH
SEA
4.
5.
Goose Hill
Kenton Hills
The Warren
Warren Cottages
Sizewell
ALDRINGHAM WITH THORPE
Sizewell Cliff
THORPENESS COMMON

11. A2727/2/11, Dunwich–Easton Bavents (sheet 62/47)

THE SEA FLOOD
JANUARY 31ST. - FEBRUARY 1ST. 1953.
MINISTRY OF AGRICULTURE AND FISHERIES.
CROWN COPYRIGHT RESERVED.
PRESENTED BY THE EAST SUFFOLK AGRICULTURAL EXECUTIVE COMMITTEE TO THE IPSWICH AND EAST SUFFOLK RECORD OFFICE.
MARCH. 6. 1954
COUNTY AGRICULTURAL OFFICER.
REFERENCE 62/47.
1. 90'.
2. 60'.
3. 120'.
4. 120'.
5. 20'.
6. 10'.
7. 2,010' SEA WASHED OVER SAND AND SHINGLE BEACH
8. 19'.
9. 50'.
10. 60'.
11. 12'.
12. 125'.
13. 158'.
14. 55'.
15. 51'.
16. 59'.
17. 1,410' SEA WASHED OVER SAND AND SHINGLE BEACH.
18. 7,500' SEA WASHED OVER SAND BEACH.
19. 30'.
20. 40'.
21. 50'.
22. 60'.
23. 40'.
24. 10'.
25. 10'.
26. 10'.
27. 20'.
28. 30'.
A2727/2/11
SOUTHWOLD
SOLE BAY
EASTON BAVENTS
Reydon
Walberswick

12. A2727/2/12, South Cove–Gisleham (sheet 62/48)

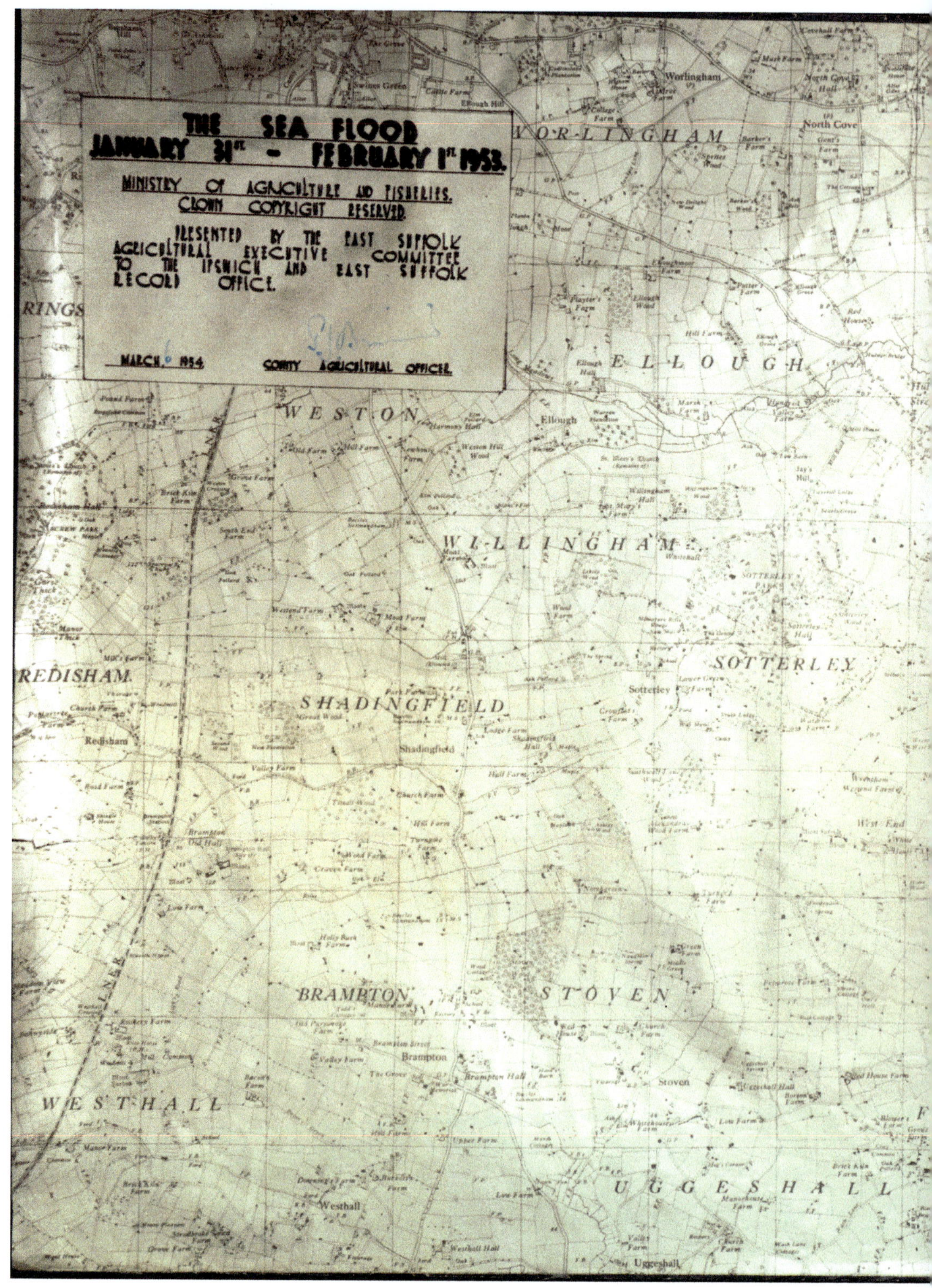

SUFF. 62/48.
Carlton Colville
Gisleham
GISLEHAM
RUSHMERE
Rushmere
Black Street
Kessingland
KESSINGLAND
Kessingland Beach
Henstead
HENSTEAD
NORTH
SEA
Benacre
BENACRE
Wrentham
Covehithe
COVEHITHE
South Cove
SOUTH COVE

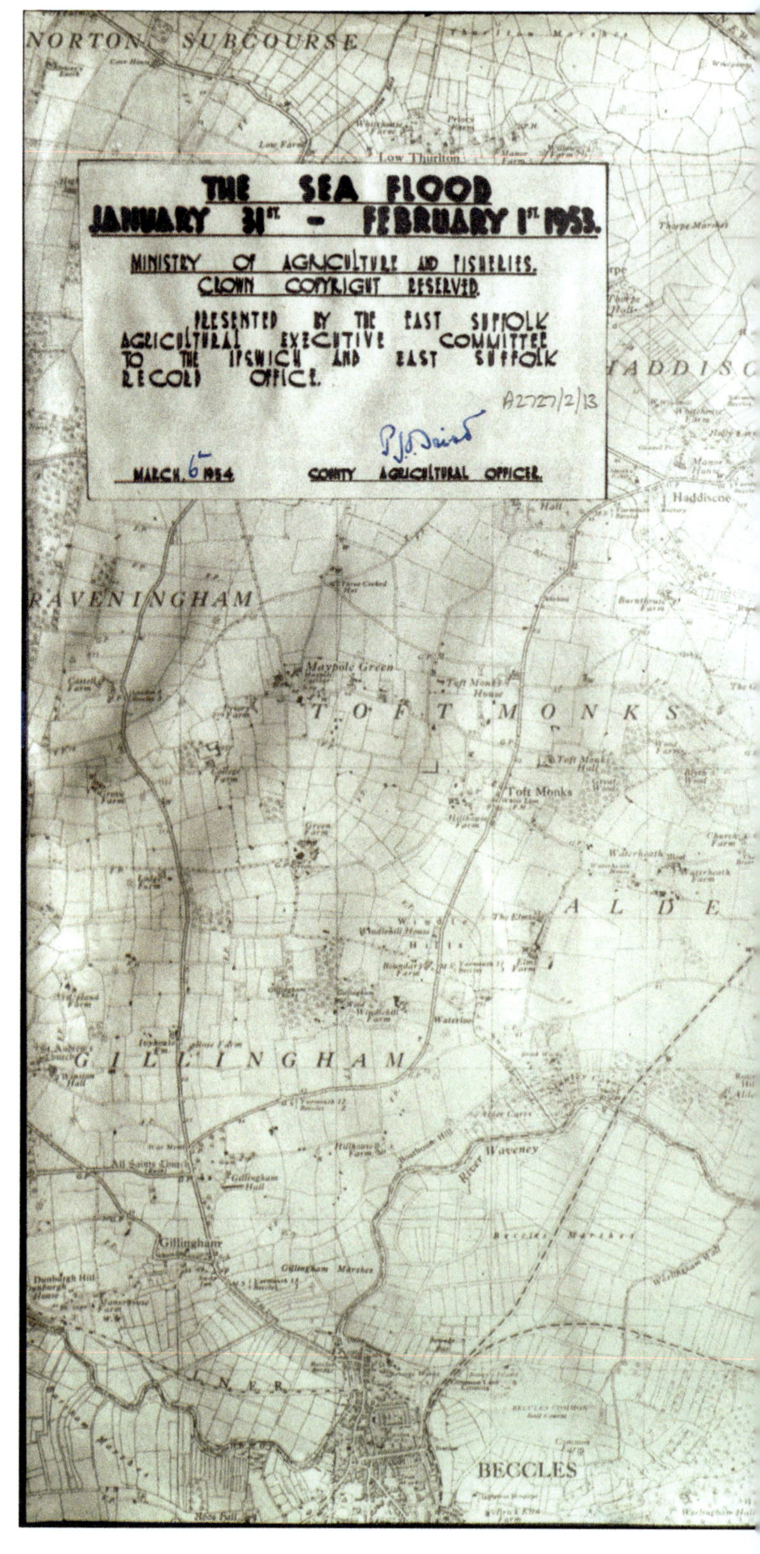

13. A2727/2/13, Barnby–Ashby (sheet 62/49)

62/49
A S H B Y
Fritton Decoy
Herringfleet Hall
H E R R I N G F L E E T
Herringfleet
Manor House Farm
River Waveney
Somerleyton Hall
SOMERLEYTON PARK
Somerleyton
S O M E R L E Y T O N
W H E A T A C R E
Wheatacre Hall
Wheatacre
B U R G H S T. P E T E R
Burgh St. Peter
Hill Farm
The Shrublands
River Waveney
B A R N B Y
L.N.E.R.
Wade Hall
NORTH COVE

14. A2727/2/14, Lowestoft–Corton (sheet 62/59)

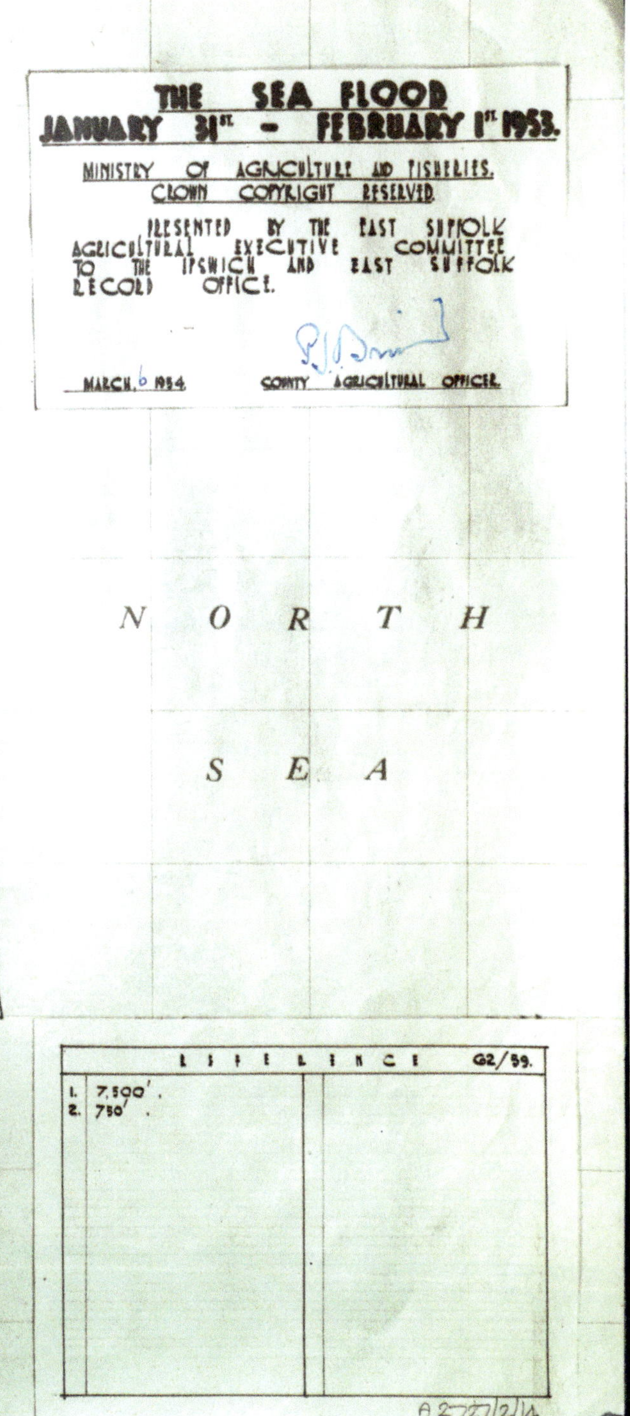
THE SEA FLOOD
JANUARY 31ST - FEBRUARY 1ST 1953.
MINISTRY OF AGRICULTURE AND FISHERIES.
CROWN COPYRIGHT RESERVED.
PRESENTED BY THE EAST SUFFOLK AGRICULTURAL EXECUTIVE COMMITTEE TO THE IPSWICH AND EAST SUFFOLK RECORD OFFICE.
MARCH 6 1954.
COUNTY AGRICULTURAL OFFICER.
N O R T H
S E A
REFERENCE G2/59.
1. 7,500'.
2. 750'.
A 2727/2/14

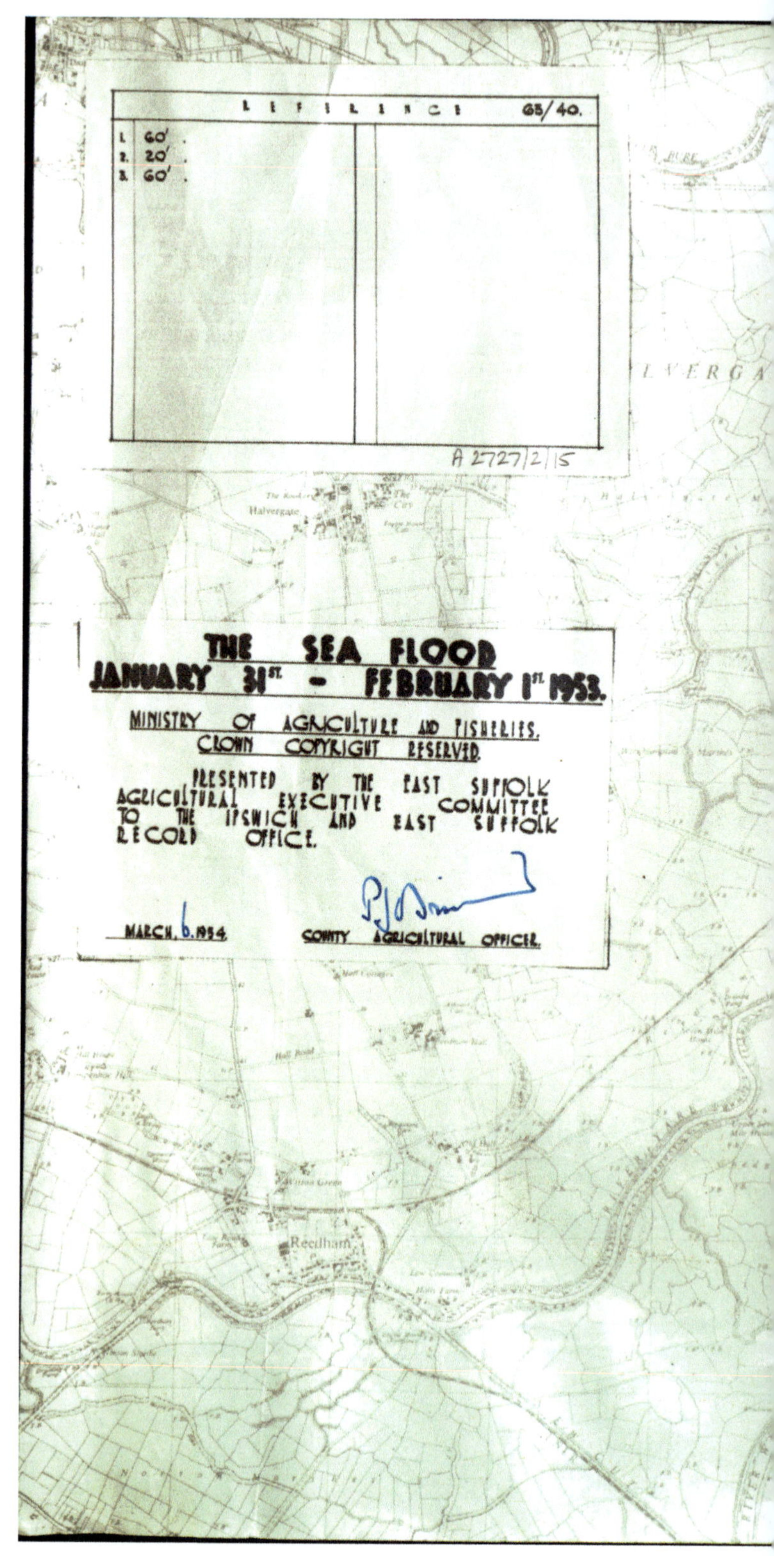

15. *A2727/2/15, Fritton–Burgh Castle (sheet 63/40)*

BURGH CASTLE
Belton
BELTON
FRITTON

16. A2727/2/16, Hopton–Great Yarmouth (sheet 63/50)

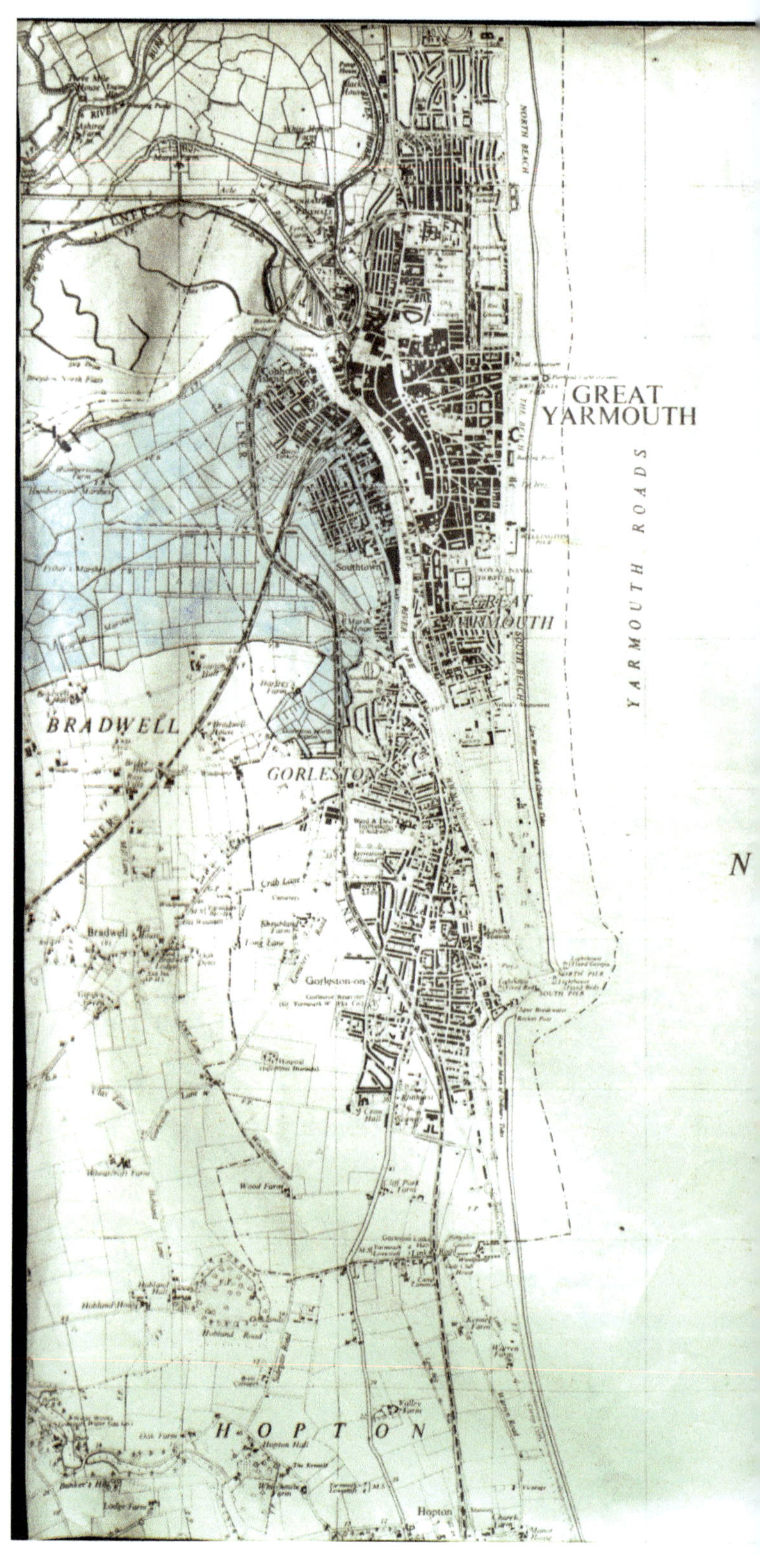

St Nicholas Lightship

SEA

THE SEA FLOOD
JANUARY 31ST. - FEBRUARY 1ST. 1953.

MINISTRY OF AGRICULTURE AND FISHERIES.

PRESENTED BY THE EAST SUFFOLK AGRICULTURAL EXECUTIVE COMMITTEE TO THE IPSWICH AND EAST SUFFOLK RECORD OFFICE.

MARCH, 1954 COUNTY AGRICULTURAL OFFICER.

REFERENCE		63/50.
1.	150'	

INDEX OF PEOPLE AND PLACES

INDEX OF SUBJECTS

THE SUFFOLK RECORDS SOCIETY

For 66 years the Suffolk Records Society has added to the knowledge of Suffolk's history by issuing an annual volume of previously unpublished manuscripts, each throwing light on some new aspect of the history of the county.

Covering 700 years and embracing letters, diaries, maps, accounts and other archives, many of them previously little known or neglected, these books have together made a major contribution to historical studies.

At the heart of this achievement lie the Society's members, all of whom share a passion for Suffolk and its history and whose support, subscriptions and donations make possible the opening up of the landscape of historical research in the area.

In exchange for this tangible support, members receive a new volume each year at a considerable saving on the retail price at which the books are then offered for sale.

Members are also welcomed to the launch of the new volume, held each year in a different and appropriate setting within the county and giving them a chance to meet and listen to some of the leading historians in their fields talking about their latest work.

For anyone with a love of history, a desire to build a library on Suffolk themes at modest cost and a wish to see historical research continue to thrive and bring new sources to the public eye in decades to come, a subscription to the Suffolk Records Society is the ideal way to make a contribution and join the company of those who give Suffolk history a future.

THE CHARTERS SERIES

To supplement the annual volumes and serve the need of medieval historians, the Charters Series was launched in 1979 with the challenge of publishing the transcribed texts of all the surviving monastic charters for the county. Since then, twenty-two volumes have been published as an occasional series, the latest in 2023.

The Charter Series is financed by a separate annual subscription leading to receipt of each volume on publication.

CURRENT PROJECTS

Volumes approved by the council of the Society for future publication include *The Incorporated Hundreds of Suffolk,* edited by John Shaw, *The Manorial Account Rolls of Walsham-le-Willows,* edited by Mark Bailey and the late Audrey McLaughlin, and *The First World War Diaries of George Punchard from Ipswich*, edited by Paul Botwright; and in the Charters Series, *Bury St Edmunds Town Charters*, two volumes edited by Vivien Brown. The order in which these and other volumes appear in print will depend on the dates of completion of editorial work.

MEMBERSHIP

Membership enquiries should be addressed to Mrs Tanya Christian, 8 Orchard Way, Needham Market, IP6 8JQ; e-mail: suffolkrecordssociety@gmail.com

The Suffolk Records Society is a registered charity, No. 1084279.

OBITUARY

ROBERT WILLIAM MALSTER 1932–2023

There is probably no one alive in Suffolk today who would be capable of naming everything that Bob Malster did for the history of his adopted county. It has also to be said that having been born into a family of hereditary freemen of the City of Norwich, he also made notable contributions to his native Norfolk with published work on its maritime past and on activity associated with its broads and rivers. Born, raised and schooled in Norwich, Bob did post-war National Service in the RAF before beginning a career in journalism with the *Eastern Daily Press,* which is what triggered his engagement with the history of 'the other county' when he was appointed *EDP* staff journalist at the office of one of its weekly local newspapers, the *Lowestoft Journal*, during the early 1950s. It was this posting which involved him in the vigorous daily life of a town dominated by North Sea fishing activity (both drifting and trawling) and by the various industries connected with it.

In due course marriage to a local girl Brenda ensued, which drew Bob into the life of the town even more, one important aspect of which was becoming part of an *ad hoc* local history group, which met weekly for lunch in Bingham's Restaurant at 89 High Street. Reference is made to this in the acknowledgements of Bob's book *Lowestoft: East Coast Port* (1982), with various people named as being influential in assisting him with the increasingly absorbing study of his place of employment.

After some years of working in Lowestoft, Bob moved on to Ipswich, to join the staff of the *East Anglian Daily Times.* This began another phase of his long connection with the county of Suffolk, helping in particular to start his specialist interest in the town and its industrial heritage. A further move, career-wise, saw him work for a number of years at the Lavenham Press as general editor for Terence Dalton Ltd, for whom he brought a number of valuable local works into print. Perhaps the most significant of these for Bob was *From Tree to Sea* (1985) by Ted Frost, a retired Lowestoft shipwright, which (along with the author's wonderful illustrations) traced the construction of the wooden Lowestoft herring drifter *Formidable* (LT 100) in 1917.

Following his eventual departure from the Lavenham Press, Bob established his own imprint, Malthouse Press, and later formed a productive relationship with Poppyland Publishing of Cromer. It was this latter connection which led to the publication of both *Maritime Norfolk*, parts one and two (2012 and 2013), and *Maritime Suffolk* (2017) – with *North Sea War 1914–1919* (2015) sandwiched in between, a considerable achievement by anyone's standards.

Bob's publications are only part of his story, however. Equally important (and perhaps even more so, on a personal level) is the depth of his generosity in the help and encouragement he gave to so many people involved in the study of East Anglian local history, be it at elementary or more advanced level. Nothing was too much trouble for him and everything he did was always carried out with that genial smile and deliciously infectious sense of humour – always accompanied by a mastery of under-statement.

He was a great user and supporter of the Ipswich branch of the Suffolk Record Office, a key figure in the Suffolk Local History Council and a valued volunteer at the Ipswich Transport Museum. He was a council member and latterly a vice-president of the Suffolk Records Society. He contributed Volume LVI (2013) to its annual series under the title of *The Minute Books of the Suffolk Humane Society*, a study which enabled him to return to one of his great loves – nineteenth-century maritime life-saving activity in the Lowestoft area.

Apart from his prolific output of publications relating to East Anglian local history, many people will remember Bob as an excellent speaker on the wide range of subjects to fall within his compass. For anyone who attended them, who can forget those Suffolk Local History Council weekend events, held at Belstead House, when Bob was one of the speakers; or, going back further, when he was on the list of WEA lecturers, travelling to all kinds of venues to create and sustain an interest in local history? His legacy is an important and varied one – and those of us who knew him will miss him.

Bob was born on 24 April 1932 in Norwich Cathedral Close and died on 16 April 2023 in Ipswich Hospital, just over a week short of his ninety-first birthday. He is survived by his daughter Andrea. May he rest easy, now that the waves have ceased to roll.

David Butcher

Printed and bound by CPI Group (UK) Ltd, Croydon, CR0 4YY

09/05/2024

14499932-0002